The Open University

Physical chemistry: principles of chemical change

Topic Study 2 Part 1
The Three-Way Catalytic Converter

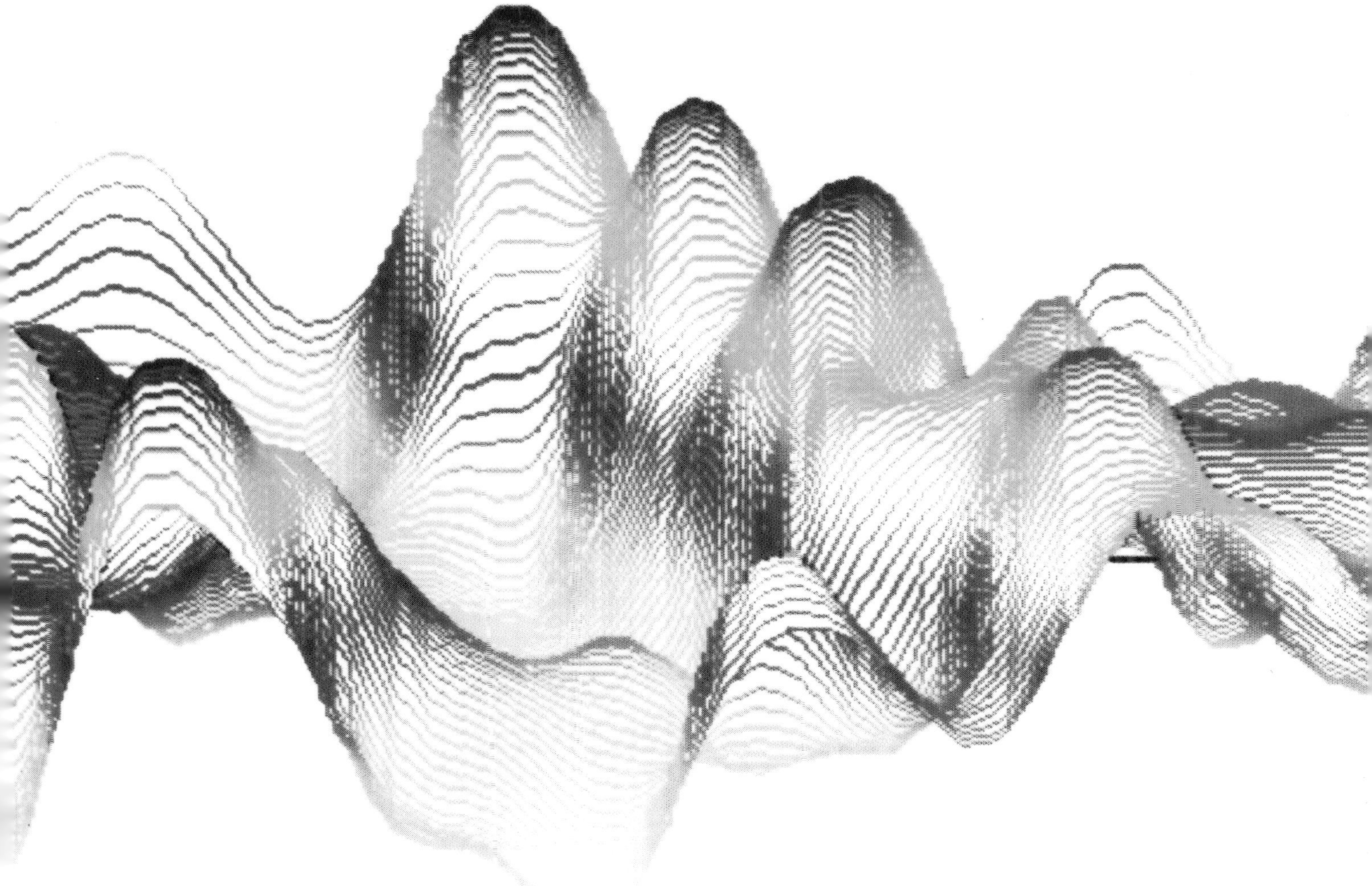

THE S342 COURSE TEAM

CHAIR AND GENERAL EDITOR
Kiki Warr

AUTHORS
Keith Bolton (Block 8; Topic Study 3)

Angela Chapman (Block 4)

Eleanor Crabb (Block 5; Topic Study 2)

Charlie Harding (Block 6; Topic Study 2)

Clive McKee (Block 6)

Michael Mortimer (Blocks 2, 3 and 5)

Kiki Warr (Blocks 1, 4, 7 and 8; Topic Study 1)

Ruth Williams (Block 3)

Other authors whose previous S342 contribution has been of considerable value in the preparation of this Course

Lesley Smart (Block 6)

Peter Taylor (Blocks 3 and 4)

Dr J. M. West (University of Sheffield, Topic Study 3)

COURSE MANAGER
Mike Bullivant

EDITORS
Ian Nuttall

Dick Sharp

BBC
David Jackson

Ian Thomas

GRAPHIC DESIGN
Debbie Crouch (Designer)

Howard Taylor (Graphic Artist)

COURSE READER
Dr Clive McKee

COURSE ASSESSOR
Professor P. G. Ashmore (original course)

Dr David Whan (revised course)

SECRETARIAL SUPPORT
Debbie Gingell (Course Secretary)

Jenny Burrage

Margaret Careford

Shirley Foster

Sue Hegarty

This publication forms part of an Open University course S342 *Physical chemistry: principles of chemical change*. Details of this and other Open University courses can be obtained from the Student Registration and Enquiry Service, The Open University, PO Box 197, Milton Keynes MK7 6BJ, United Kingdom: tel. +44 (0)845 300 60 90, email general-enquiries@open.ac.uk

Alternatively, you may visit the Open University website at http://www.open.ac.uk where you can learn more about the wide range of courses and packs offered at all levels by The Open University.

To purchase a selection of Open University course materials visit http://www.ouw.co.uk, or contact Open University Worldwide, Walton Hall, Milton Keynes MK7 6AA, United Kingdom for a brochure. tel. +44 (0)1908 858793; fax +44 (0)1908 858787; email ouw-customer-services@open.ac.uk

The Open University
Walton Hall, Milton Keynes
MK7 6AA

First published 1996. Reprinted 2008.

Edited and designed by The Open University.

Typeset by The Open University.

Printed in the United Kingdom by Martins the Printers, Berwick-upon-Tweed.

ISBN 978 0 7492 5191 8

1.3

CONTENTS

1 INTRODUCTION

The dreadful urban smogs or 'pea soupers' of the 1950s are still within the living memories of many people. One such smog is thought to have led to the premature deaths of 4 000 people in London during the winter of 1952. These conditions were mainly caused by the smoke and sulfur dioxide generated by burning coal: they were all but eliminated by a combination of measures, of which the most important were the Clean Air Acts of 1956 and 1968, and the introduction of smokeless zones.

More recently, however, the quality of our air is once more a matter of public concern, particularly in urban areas, largely due to emissions of pollutants from motor traffic. This concern, and increased public awareness of the importance of air quality, have led the UK Government to establish a public information system that issues daily bulletins on air quality through the media (Figure 1). Although the current situation is markedly less serious than the smogs of the 1950s, pollutant concentrations regularly exceed international health guidelines in our towns and cities. On 13 December 1991, London experienced the highest levels of nitrogen dioxide (NO_2) since regular monitoring began in 1972.

The UK Government and the European Union are responding to this problem with a range of pollution abatement measures designed to cut emissions from motor traffic, of which the three-way catalytic converter is probably the best known. Here, we shall look at the development of emission control catalysts for petrol-driven vehicles and examine the chemistry involved.

Figure 1   Weather map showing air quality forecast for the UK.

STUDY COMMENT   This part of Topic Study 2 builds on principles and methods discussed in Blocks 5 and 6. In particular, it discusses the development of catalysts to control vehicle emissions from both a scientific and a historical perspective. It will also look at the information that surface-science techniques have provided about the mechanisms of catalytic reactions, and, perhaps more importantly from an industrial point of view, about the problem of catalyst deactivation. Band 7 on videocassette 2 (*A clean get-away!*) provides an insight into research being carried out by Johnson Matthey, major manufacturers of the catalytic converter, aimed at improving the efficiency of the current system. This sequence is probably best viewed when you have finished studying Section 6.

Finally, you will find, as in Topic Study 1, that the material is generally more descriptive than that in the main Blocks, and that boxes are again used to expand on points in the text, or to provide extra background information.

2 EXHAUST POLLUTANTS:
SOURCES AND EFFECTS

Before we start our discussion of emission control catalysts, we need to take a brief look at both the **primary pollutants** (the gases emitted directly from vehicle exhausts), and at the products to which they can give rise once released into the atmosphere – the **secondary pollutants**.

2.1 Primary pollutants

The most important chemical reaction in a petrol engine – that is, the one that provides the energy to drive the vehicle (see Box 1) – is the combustion of fuel in air. Petrol can contain well over 300 hydrocarbons, with carbon numbers ranging from C_4 to C_{12}, and with boiling temperatures ranging from 25 °C to 220 °C, as well as a variety of additives such as detergents and anti-oxidants. The *key* constituents are listed in Table 1. Taking octane, C_8H_{18}, as a typical constituent, then in an 'ideal' system, combustion would be complete, with carbon dioxide and steam as the only products:

$$C_8H_{18}(g) + 12\tfrac{1}{2}O_2(g) = 8CO_2(g) + 9H_2O(g) \tag{1}$$

Table 1   The key constituents of petrol.

Constituents	Quality of fuel	Approximate 'winter' composition[a]/vol %
straight run gasoline the gasoline fraction from crude oil, containing mainly saturated C_4–C_7	low	5
catalytic reformate produced by converting *n*-alkanes into *iso*-alkanes and cycloalkanes into aromatics	good	40
cracked spirit includes many unsaturated compounds formed by splitting up higher fractions	moderate	30
alkylate couples small alkanes and alkenes to produce *iso*-octane, for example	excellent	10
oxygenates compounds such as ethers and alcohols; for example, methyltertiarybutyl ether (MTBE), which is often used to replace leaded compounds	very good	5
butane added to increase the volatility of petrol, to allow vehicles to start in cold weather	very good	8

[a] Refineries deliver slightly different grades of fuel in summer and winter – volatility, in particular, will vary with the seasons.

BOX 1

THE COMBUSTION ENGINE

Although the petrol-driven motor car typically runs on four cylinders, the essential features of its operation can be understood by considering the sequence of events in a single cylinder, shown in a simplified form in the Figure. The conventional cylinder employs a four-stroke cycle. As the piston descends, it draws in an air/fuel vapour mixture (through the inlet valve). The inlet valve then closes, and the piston ascends to the top of its stroke, compressing the air and fuel mixture. Just before the moment of maximum compression, a spark from the plug ignites the mixture; the exothermic reaction causes a sudden temperature rise, accompanied by rapid expansion of the gases in the cylinder, which propels the piston downwards. This provides the energy to turn the crankshaft, which in turn drives the wheels of the car. Gas temperatures during the combustion are typically in excess of 1 500 °C. Finally, the exhaust valve opens as the piston returns to the top of its stroke, and the combustion products are driven into the exhaust system at temperatures of 900–1 000 °C. The gases cool down as they pass through the exhaust, and exit into the atmosphere at about 200 °C.

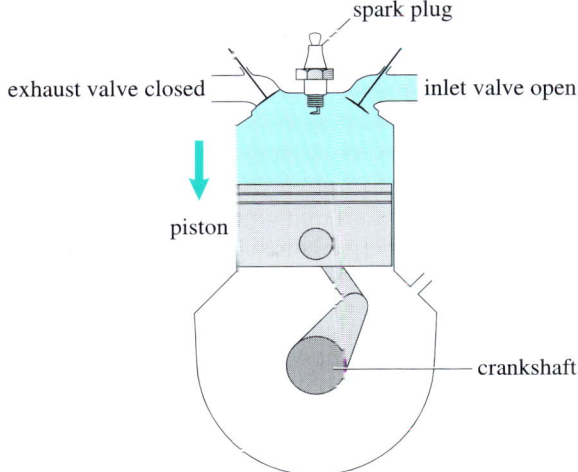

A schematic view of one cylinder of an internal combustion engine at the point when the air/fuel mixture is entering through the inlet valve.

Complete oxidation of the fuel depends on a number of factors: first, there must be sufficient oxygen present; second, there must be adequate mixing of the petrol and air; and finally, there must be sufficient time for the mixture to react at high temperature before the gases are cooled. In internal combustion engines, the time available for combustion is limited by the engine's cycle to just a few milliseconds. In practice, *incomplete* combustion of the fuel leads to emissions of the partial oxidation product, carbon monoxide (CO), and a wide range of **volatile organic compounds** (VOC), including **hydrocarbons** (HC), aromatics and oxygenated species. These emissions are particularly high during both idling and deceleration, when insufficient air is taken in for complete combustion to occur (Figure 2).

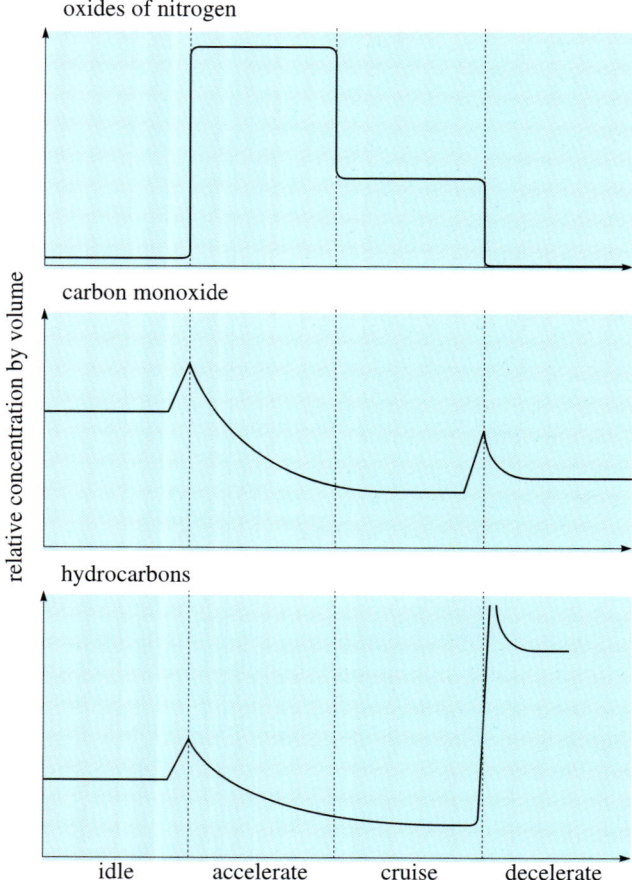

Figure 2 Variation in exhaust emissions with driving mode.

Another important result of the combustion process, particularly during acceleration, is the production of the oxides of nitrogen – nitric oxide (nitrogen monoxide, NO) and nitrogen dioxide (NO_2) – conventionally considered together, and represented as NO_x. At the high temperatures involved (in excess of 1 500 °C), nitrogen and oxygen in the air drawn in with the fuel, may combine to form nitric oxide:

$$N_2(g) + O_2(g) = 2NO(g) \tag{2}$$

On leaving the exhaust system, NO cools down and is oxidized by oxidants in the atmosphere to form the dioxide, NO_2 (see Section 2.2). Although the 'fixing' of nitrogen from the air, reaction 2, is the major source of NO_x, it may also arise from the oxidation of any nitrogenous components in the fuel.

The information collected in Table 2 emphasizes just how significant road transport is as a source of air pollutants. The scale of the problem is further emphasized by the data for the emissions from just one medium-sized (2 litre) petrol engine, without a catalyst: for each 10 000 miles it travels, it produces 700 kg of CO, 150 kg of HC and 130 kg of NO_x!

Table 2 Contributions of different sources of the principal primary air pollutants (excluding CO_2) in the UK in 1990. The figures have been rounded up to the nearest whole number, so the percentage totals do not always add up to 100.

	% of total emissions				
Source	SO_2	NO_x	CO	VOC	Black smoke
road transport	2	51	90	41	46
electricity supply industry	72	28	1	–	6
other industry	19	9	4	52	14
domestic	3	2	4	2	33
other	7	9	–	4	1
total / kilotonnes (kT) [a]	3 774	2 719	6 659	2 396	453

[a] 1 tonne = 10^3 kg.

2.1.1 Harmful effects of primary exhaust emissions

None of the primary pollutants is a desirable addition to the atmosphere. Carbon dioxide is harmless to humans at concentrations below about 1% by volume. As was noted in Topic Study 1, however, there are growing concerns over its wider role as a greenhouse gas, and the possible climatic effects of the ongoing accumulation of CO_2 in the atmosphere. By contrast, CO is a very poisonous gas, interfering with the transfer of oxygen through the body by displacing oxygen in the haemoglobin of red blood cells. NO is relatively innocuous but it is oxidized to NO_2, which can cause breathing difficulties, particularly in young children or asthmatics. On 13 December 1991, and during other pollution 'episodes', a large number of people reported severe respiratory problems. NO_x can also undergo further reaction in the atmosphere to form dilute nitric acid, which is thought to be a major contributor to acid rain.

Globally, amounts of NO_x produced naturally (by bacterial and volcanic action and by lightning) far outweigh anthropogenic emissions. These natural sources give rise to a low background level of NO_x in the atmosphere. Anthropogenic emissions add to this background level, but the highest concentrations tend to be localized relatively close to the source. In fact, NO_2 levels rarely exceed guidelines at distances greater than 10 m from the roadside. Thus, high NO_x levels are of particular concern in urban areas.

Some of the unburned hydrocarbons released as VOC, especially benzene and other aromatics, as well as their partially oxidized products (such as polynuclear aromatic hydrocarbons, PAHs), are believed to have deleterious effects on health, and in particular to have possible links with cancer. But the *major* effect of the hydrocarbons is their role in the formation of secondary pollutants.

2.2 Secondary pollutants

Perhaps the most notorious consequence of exhaust emissions is their role in the formation of **photochemical smog**, first detected in Los Angeles in the 1940s. Photochemical smog, evident in Figure 3 as a (brown) haze, is a mixture of ozone, nitrogen dioxide and other secondary products, together with small particles. The resulting poor visibility can itself be a hazard, and ozone attacks rubber and other organic materials. Ozone, nitrogen dioxide, and especially the particles, are irritants to the respiratory system and can cause severe damage to human health. Plant growth may also be impaired.

There are several factors that cause the problem to be so severe in Los Angeles. The city is enclosed on three sides by high ground, and this often results in lighter, warm air lying on top of heavier, cool air – a very stable situation known as a 'temperature inversion' (Topic Study 1) – which leads to the trapping of pollutants. Intensely

Figure 3   Photographs of Los Angeles (left) in clean air and (right) after smog has formed.

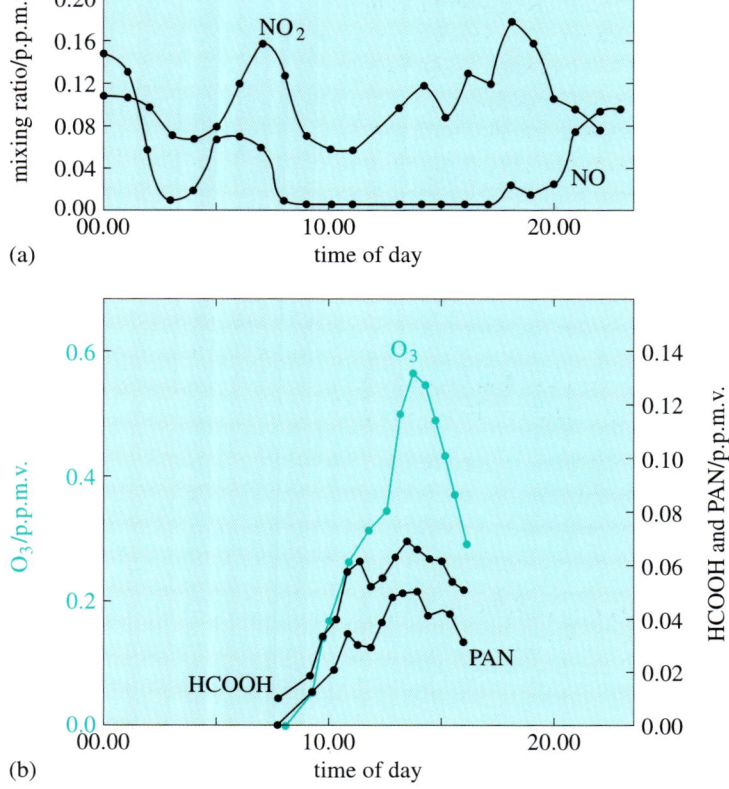

(a)

(b)

Figure 4 Variations in the concentration of some (a) primary and (b) secondary pollutants during the course of a smoggy day in Southern California. (HCOOH is methanoic acid, and PAN stands for peroxyacetyl nitrate.)

sunny days are frequent, and there is a very high density of vehicle use. Mixed with air, the primary pollutants in the exhaust fumes (CO, NO_x and hydrocarbons) undergo chemical change on irradiation with sunlight, giving rise to the concentration changes and secondary pollutants shown in Figure 4.

Ozone is present in the troposphere at a *background* level of 20–50 p.p.b.v. (0.02–0.05 p.p.m.v.), mainly as a result of downward transport from the stratosphere. Ozone builds up to the much higher levels evident in Figure 4b because it can also be generated *within* the lower atmosphere, again via the reaction:

$$O + O_2 + M \longrightarrow O_3 + M \tag{3}$$

where M is a third body, typically N_2 or O_2.

As you saw in Topic Study 1, photolysis of O_2 provides the source of atomic oxygen needed for ozone production in the stratosphere. But the solar radiation penetrating to near ground level is virtually all at wavelengths greater than about 300 nm. Relatively few chemical species can be photolysed at these wavelengths to give oxygen atoms, and the only known candidate is NO_2:

$$NO_2 + h\nu \longrightarrow NO + O \qquad (\lambda < 430\,nm) \tag{4}$$

Since NO, rather than NO_2, is the primary pollutant in vehicle exhausts, there must be some way of converting NO into NO_2. The reaction with molecular oxygen:

$$2NO + O_2 \longrightarrow 2NO_2 \tag{5}$$

is too slow to be important at the relatively low concentrations of nitric oxide found in polluted air (its rate depends on $[NO]^2$). In the absence of hydrocarbons and CO, the only process that is fast enough is reaction with ozone:

$$NO + O_3 \longrightarrow NO_2 + O_2 \tag{6}$$

Since ozone is needed to effect this conversion, and NO_2 is necessary for ozone formation, it is apparent that the limited scheme given in reactions 3, 4 and 6 cannot, *by itself*, explain the build up of O_3 evident in Figure 4b. Rather, it represents a closed cycle of reactions (captured in Figure 5), whereby a dynamic balance is established between ozone destruction and formation. *Net* ozone production requires an alternative route for converting NO into NO_2 – one that does *not* consume O_3. The presence of hydrocarbons, and to a lesser extent CO, in the exhaust gas cocktail provides just such a route.

The key species involved is the hydroxyl radical HO· (recall Box 1 in Topic Study 1). Attack by HO· abstracts a hydrogen atom from a hydrocarbon (RCH_3, say):

$$RCH_3 + HO· \longrightarrow RCH_2· + H_2O \qquad (7)$$

where R can be H or an alkyl group. The alkyl radical ($RCH_2·$) can add O_2 to form a peroxyalkyl radical ($RCH_2O_2·$), which reacts with NO to generate an alkoxy radical ($RCH_2O·$):

$$RCH_2· + O_2 + M \longrightarrow RCH_2O_2· + M \qquad (8)$$

$$RCH_2O_2· + NO \longrightarrow RCH_2O· + NO_2 \qquad (9)$$

Further reaction with molecular oxygen yields an aldehyde (alkanal), RCHO, and a hydroperoxy radical ($HO_2·$). The latter can, in turn, react with NO to regenerate HO·:

$$RCH_2O· + O_2 \longrightarrow RCHO + HO_2· \qquad (10)$$

$$HO_2· + NO \longrightarrow HO· + NO_2 \qquad (11)$$

■ What is the net effect of the elementary steps in reactions 7-11?

▨ Adding these steps together gives the following overall equation:

$$RCH_3 + 2NO + 2O_2 = RCHO + 2NO_2 + H_2O \qquad (12)$$

In short, reactions 7–11 provide a mechanism for oxidizing hydrocarbons, *while at the same time converting* NO *into* NO_2. Assembling the steps as a cycle (Figure 6) highlights the feature – the regeneration of HO· (reaction 11) – that allows the sequence to be repeated many times over. Each trip around the cycle produces two molecules of NO_2, and hence, potentially at least, two molecules of O_3 – via reactions 4 and 3.

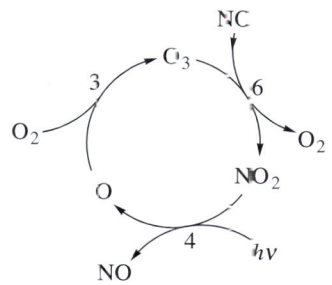

Figure 5   Cycle showing the destruction/regeneration of O_3 and NO_2 in the lower atmosphere via reactions 6, 4 and 3.

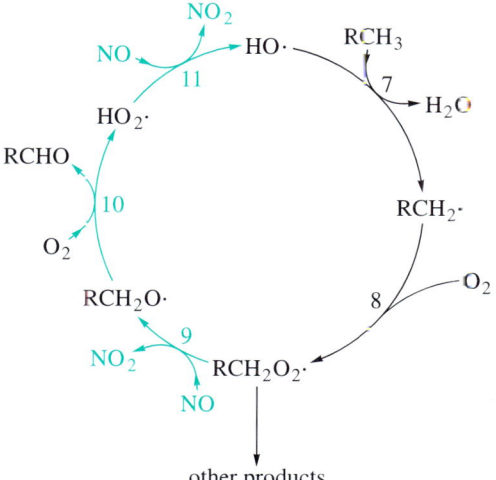

Figure 6   The essential steps (reactions 7–11) in tropospheric hydrocarbon oxidation. The steps shown in black can occur in NO-poor air. In NO-rich environments, the processes shown in green can close a cycle, with regeneration of HO·, conversion of NO into NO_2, and hence *net* ozone production.

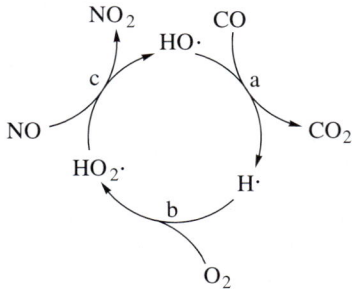

Figure 7 The essential steps in the tropospheric oxidation of CO, in the presence of NO.

SAQ 1 (revision) Figure 7 shows an analogous cycle for the tropospheric oxidation of CO in the presence of NO. By writing an equation for each of the steps in the cycle, and then including the photolysis of NO_2 (equation 4) and the formation of O_3 (equation 3), show that the overall effect can be represented as follows:

$$CO + 2O_2 + hv \longrightarrow CO_2 + O_3 \tag{13}$$

The schemes in Figures 6 and 7 are sufficient for our purposes: they show how sharply enhanced ozone concentrations can arise when suitable amounts of *both* NO_x *and* hydrocarbons (and/or CO) are present. But you should bear in mind that the situation in the 'real' atmosphere is far more complicated. As you might expect, competitive processes abound. In particular, the HO· radical is a key participant in many other reaction sequences as well, several of which can, in turn, influence ozone concentrations. And of course, meteorological factors (wind speed, temperature, the amount of solar radiation, etc.) will also come into the equation. Computer modelling studies are the only way of simulating the combined effect of all of these factors – just as they are in the context of stratospheric chemistry. Suffice it to say here that such studies can reproduce important features of pollution 'episodes' in the real world. One telling example is the fact that the maximum ozone concentration often occurs downwind of the source of primary pollutants, so that, typically, it is found in rural rather than urban areas. It has been estimated that this has cost the USA considerably in terms of damage to crops.

Chief among other chemicals found in photochemical smog are a variety of oxygenated organic compounds such as aldehydes, carboxylic acids (note the presence of methanoic acid, HCOOH, in Figure 4, for example) and peroxyacyl nitrates. The latter are powerful *lachrymators,* causing intense irritation of the eyes, nose and throat. They are also harmful to plants, causing leaf-damage in particular. The most common member of this group is peroxyacetyl nitrate (PAN, also in Figure 4), which can be formed from acetaldehyde (ethanal) as follows:

$$CH_3CHO + HO \cdot \longrightarrow CH_3CO \cdot + H_2O \tag{14}$$

$$CH_3CO \cdot + O_2 \longrightarrow CH_3\overset{\overset{O}{\|}}{C}OO \cdot \tag{15}$$

$$CH_3\overset{\overset{O}{\|}}{C}OO \cdot + NO_2 \rightleftharpoons CH_3\overset{\overset{O}{\|}}{C}OONO_2 \tag{16}$$

$$\text{PAN}$$

Another undesirable effect of photochemical smog is the reduced visibility that arises because light is absorbed and scattered by fine airborne particles (suspended particulate matter) or *aerosols*. Organic aerosols include aliphatic, aromatic and oxygenated compounds, including PAN. Inorganic aerosols include aqueous droplets containing sulfate, nitrate and ammonium ions, as well as a variety of trace metal particles. Sulfate and nitrate aerosols are produced via the photochemical oxidation of SO_2 and NO_x, respectively, but the detailed chemistry involved is complex, and still not fully understood. We shall not pursue the matter any further.

As noted earlier, pollutant concentrations, and hence photochemical smog formation, depend strongly on meteorological conditions. Formation is favoured by limited air circulation, and is commonly found when an area experiences a temperature inversion. The local topography clearly has an important role to play in this – witness the problem in Los Angeles. But temperature inversions can occur anywhere, particularly during overnight cooling of the ground – especially under anticyclonic (high pressure) conditions (see Box 2).

═ BOX 2 ═

METEOROLOGICAL INFLUENCES ON AIR QUALITY

The Figure shows a weather map for 13 December 1991. The thin lines represent isobars, which are used to join areas of equal pressure. 'H' represents a region of high pressure or an anticyclone; 'L' represents a low pressure region, depression or cyclone.

In a region of high pressure, skies are usually clear, with plenty of sunshine during the day. In summer, this usually makes the weather pleasantly warm, but it often produces crisp cold weather in winter. Under these conditions, clear skies at night allow the ground to cool rapidly, and a temperature inversion may then occur. And that can set the scene for a major pollution 'episode' during the winter months: cold weather stimulates combustion-generated air pollution which may stagnate under the stable conditions associated with an anticyclone; strong sunshine during daylight hours can then generate secondary pollutants, which are also trapped under the inversion. Most of the major episodes, including the notorious smogs of the 1950s, occurred during

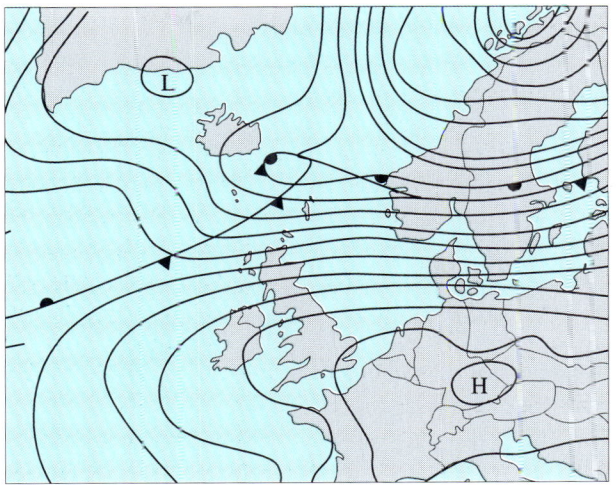

such anticyclones; indeed, as the Figure shows, Britain and Europe were experiencing a band of high pressure on 13 December 1991 – leading to the worst NO_x episode in London since monitoring began in 1972.

2.3   Air quality guidelines and environmental monitoring

Limits and guidelines for maximum concentrations of pollutants have been set by the European Union (EU) and the World Health Organisation (WHO), among various other bodies. The guidelines are typically expressed as a maximum volume mixing ratio, p.p.m.v. (or p.p.b.v.), or mass per unit volume, mg m^{-3} (or μg m^{-3}). They are usually related to specific periods of time; for example, WHO has proposed the following time-weighted exposure guidelines for CO:

- 100 mg m^{-3} (87 p.p.m.v.) over 15 minutes;

- 60 mg m^{-3} (50 p.p.m.v.) over 30 minutes;

- 30 mg m^{-3} (25 p.p.m.v.) over 1 hour;

- 10 mg m^{-3} (10 p.p.m.v.) over 8 hours.

For NO_x, the EU has set a limit of 105 p.p.b.v. (200 μg m^{-3}) in terms of the 98th percentile hourly average over the year. This means that the pollutant concentration must be below this level for 98% of the hourly averages in a year; that is, it must not exceed this level for more than 175 hours (2%) in a year. During the pollution episode in December 1991, NO_x values were recorded in London at 382 p.p.b.v. (730 μg m^{-3}), but levels were below the limit for most of the year, and the 98th percentile was not exceeded.

SAQ 2 (revision)   Concentrate on the limit for the 98th percentile hourly average for NO_x noted above. Show that the values are equivalent to one another; that is, 105 p.p.b.v. is equivalent to 200 μg m^{-3}. Assume the number density of air at ground level has the value quoted in Topic Study 1, $[M] = 2.5 \times 10^{19}$ (molecules) cm^{-3} or strictly, $[M] = 2.5 \times 10^{19}$ cm^{-3}, and take the molar mass of NO_x to be that of NO_2 (46.0 g mol^{-1}).

The UK Department of the Environment air quality guidelines are given in Table 3. SO_2, NO_2, O_3, CO and particulates are monitored continuously at 24 sites in the major cities in the UK on the Enhanced Urban Network. The Rural Air Quality Network consists of 17 sites monitoring ozone, with three sites additionally monitoring SO_2 and NO_x. In addition, over 300 local authorities in the UK participate in a network to monitor NO_2, measuring monthly NO_2 means at a kerbside site, at a site 20–30 m from a major road, and at two rural background sites.

There are basically two types of equipment for monitoring air quality:

- Continuous automatic equipment, using sophisticated techniques to give more or less instantaneous measurements, allowing peak concentrations to be measured.

- Simple, less sensitive, techniques which give averages over longer times, typically 24 hours to 1 month. In recent years, diffusion tubes specific to one pollutant have been developed. They have been widely used to measure ambient levels of NO_2 and are used in the nationwide monitoring network.

Data collected by the networks are made available via an air quality telephone line, the World Wide Web and weather bulletins.

Table 3   The Department of the Environment air quality guidelines expressed as 1 hour means in p.p.b.v.

	Nitrogen dioxide	Sulfur dioxide	Ozone
very good	<50	<60	<50
good	50–99	60–124	50–89
poor	100–299	125–399	90–179
very poor	⩾300	⩾400	⩾180

2.4   Summary of Section 2

1   The three classes of primary pollutants emitted from a petrol-driven vehicle (in addition to H_2O and CO_2) are: carbon monoxide and volatile organic compounds (VOC), including hydrocarbons (HC), produced from incomplete oxidation; and the nitrogen oxides, NO_x, produced by fixing nitrogen from the air at the high temperatures (in excess of 1 500 °C) involved in the combustion reaction. The gases are emitted from the engine into the exhaust at temperatures in the range 900–1 000 °C, cooling down as they pass through the exhaust, until they exit at about 200 °C.

2   None of these three classes of pollutants is a desirable addition to the atmosphere, causing problems for health and contributing to environmental hazards such as acid rain.

3   A major consequence of the mixture of exhaust gases is that they provide the necessary ingredients for the formation of secondary pollutants, such as ozone, peroxyacetyl nitrate (PAN) and aerosols, found together in photochemical smog.

3   LEGISLATION AND EMISSION TESTING

3.1   Legislation

The problem of photochemical smog was first related to the photochemical interactions of NO_x, hydrocarbons and oxygen in the early 1950s. Surveys established that emissions from motor vehicles were a major factor in the build-up of NO_x and hydrocarbons. The obvious way of alleviating the problem of photochemical smog would be to eliminate, or at least reduce, the primary emissions from vehicles. There are various possible approaches. An ideal solution would be to stop them forming in the first place; however, this is not possible with current petrol-driven vehicles. Alternative fuels such as hydrogen or methanol would eliminate at least some of the problem. Modification of the engine can also help to a limited extent. Another approach is to destroy the pollutants before they exit from the exhaust pipe. This is where catalytic converters come into play. As you will see in Section 4, catalyst development has been driven by the changes in emission legislation outlined below.

The state of California was the first to introduce automotive emission regulations in 1970, in an effort to reduce the problem of poor air quality. Similar legislation was quickly adopted by the US Federal Government in its 1970 amendments to the US Clean Air Act. Since then, progressively more stringent limits have been set in the USA for emissions of hydrocarbons, NO_x and CO, as shown in Figure 8. Standards already set in California until the year 2003 are shown in Table 4. Current limits (1994) represent a reduction from uncontrolled levels of over 90% for all three pollutants. Standards have also been set for commercial and diesel-fuelled vehicles, but space does not allow us to consider them in this Topic Study.

Figure 8   Changes in the US Federal emission standards for maximum limits in grams per mile of hydrocarbons, NO_x and CO, for new vehicles with respect to model year.

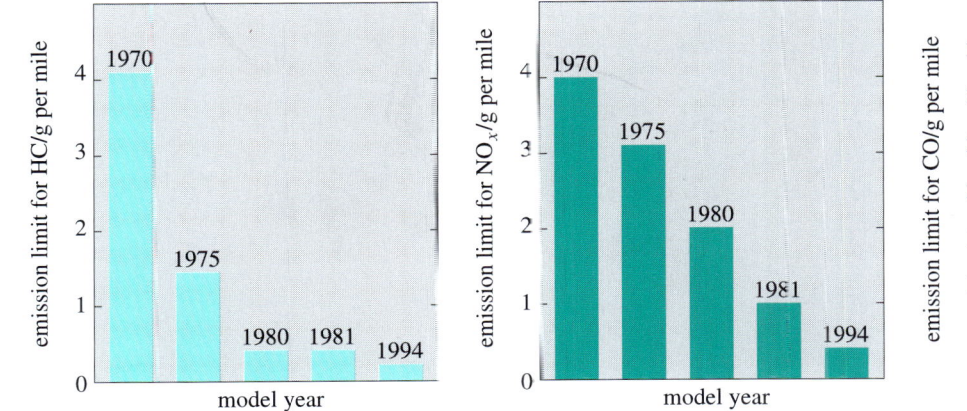

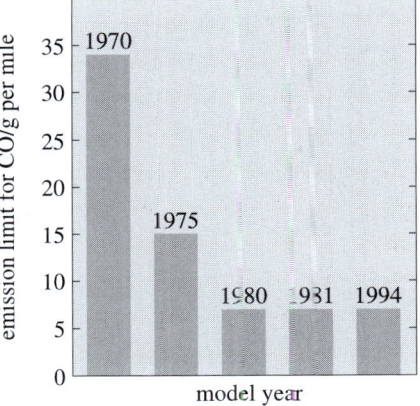

Table 4   Californian passenger car emission standards in grams per mile in 1994 and beyond, together with the percentage of new vehicles that must comply with the standards in each year.

	TLEV [a]	LEV [b]	ULEV [c]	ZEV [d]
hydrocarbons	0.125	0.075	0.040	0.0
CO	3.4	3.4	1.7	0.0
NO_x	0.4	0.2	0.2	0.0
Implementation year				
1994	10%			
1995	15%			
1996	20%			
1997		25%	2%	
1998		48%	2%	2%[e]
1999		73%	2%	2%[e]
2000		96%	2%	2%[e]
2001		90%	5%	5%[e]
2002		85%	10%	5%[e]
2003		75%	15%	10%

[a] Transitional low emission vehicle; [b] Low emission vehicle; [c] Ultra low emission vehicle; [d] Zero emission vehicle. [e] At the time of writing, it seems probable that these intermediate targets will be dropped, at the request of the car-makers, because suitable technology has not yet been developed.

Similar, if generally less stringent, emission standards have now been adopted throughout the industrialized world. The changes in European legislation for new vehicles are shown in Figure 9; the most recent standards at the time of writing come into effect for new car models from 1 January 1996, and for all new cars from 1997.

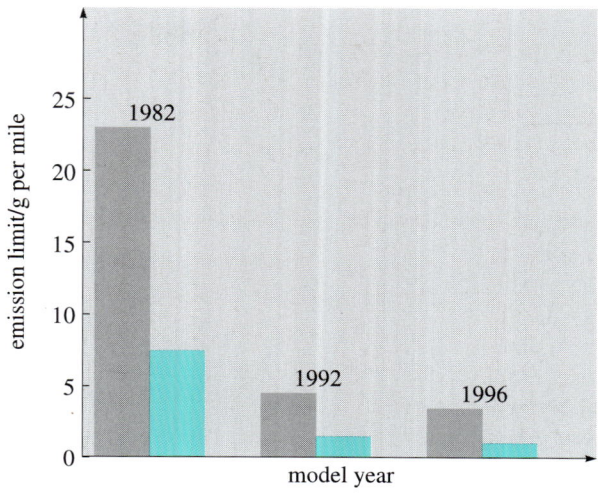

Figure 9 Changes in the EU Directive limits for maximum CO (grey) and combined HC + NO$_x$ (green) emission levels in grams per mile for new light-duty vehicles (petrol-fuelled) for different model years. These are the levels with which manufacturers must comply before they are allowed to introduce a new vehicle on to the market.

In addition, in the UK, *in-service* emissions testing has been part of the annual MOT test since November 1991, and roadside testing is also becoming more commonplace. These are the standards with which all vehicles on the road must comply. (The standards currently enforced for petrol-engined cars registered before 1 August 1992 are outlined in Box 3. *All* petrol-engined cars registered after this date (catalytic converters became mandatory for new models from this date and for all new cars from 1 January 1993) must go to a specially approved 'catalysts tested' MOT station, where more stringent, vehicle-specific, standards apply.)

=== **BOX 3** ===

MOT EMISSION STANDARDS (25 SEPTEMBER 1995)

- For vehicles first used on or after 1 August 1986: a maximum of 3.5% CO (by volume) in the exhaust gas;

- For vehicles first used between 1 August 1975 and 31 August 1986: a maximum of 4.5% CO in the exhaust gas;

- For vehicles first used on or after 1 August 1975: a maximum of 1 200 p.p.m.v. of hydrocarbons in the exhaust gas;

- For all vehicles: a check that there is not excessive smoke from the exhaust;

- For all vehicles: a general proviso that CO emission levels will not be required to be reduced below the vehicle manufacturer's specification for the engine fitted to the vehicle.

3.2   Emission level test cycles

To ensure that new vehicle models comply with the latest emission legislation, it is necessary to define a rigorous and highly reproducible test cycle over which exhaust gas compositions can be measured. For example, the European Test Procedure (which is based on the US Federal Test Procedure) is designed to mimic urban driving conditions, in which the engine starts from cold and is then subjected to repeated stop–start driving (Figure 10). The duration of the test, the distance 'travelled', and each engine mode (acceleration, deceleration, idle, and cruise) are exactly prescribed. The test is carried out by placing the vehicle on a chassis dynamometer, which continuously monitors the apparent speed of the car. All the gases emitted from the exhaust-pipe are collected in bags of uniform volume, with each bag corresponding to a specific portion of the test procedure, thus allowing gas analysis to be related to particular phases in the engine cycle. The total mass of the pollutant emitted during the test is averaged over the total distance 'travelled', to give an emission value expressed in grams per mile. New vehicles must meet the current standard before they will be approved for sale.

Although the European test is based on the US Federal Test Procedure, the 'driving cycle' is significantly different. The maximum and average speeds are lower than in the USA (31.0 and 11.6 mph, compared with 56.5 and 21.1 mph), with the engine idle during 31% of the test to simulate city driving in congested conditions. A phase to simulate motorway driving has recently been added, however, and within this Extra Urban Driving Cycle (EUDC) the speed reaches 75 mph (Figure 10).

These tests are used to compare different catalysts, and to test their response to various fuel compositions and **ageing** schedules (exhaust gas composition, temperature, time, etc.). They can also be used to examine the *transient conversion efficiency*; that is, the time required for the catalytic converter initially to start functioning.

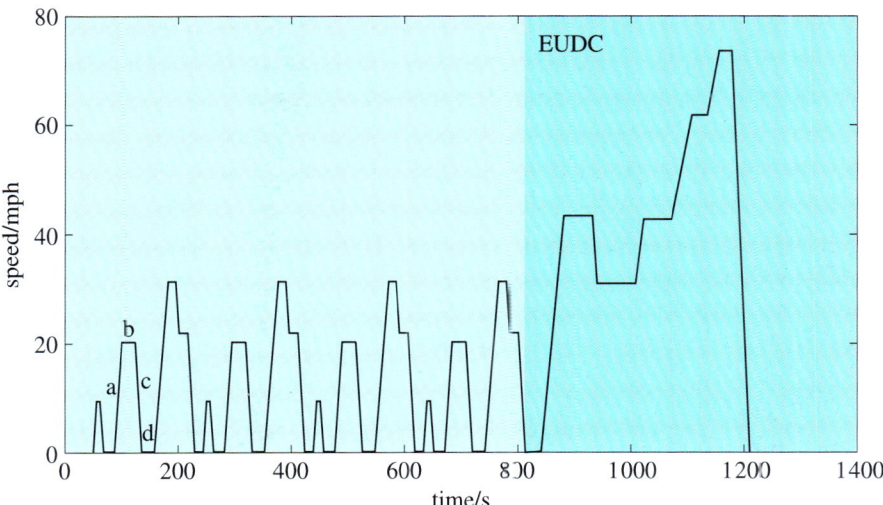

Figure 10   The European Test procedure, including the Extra Urban Driving Cycle (EUDC). The annotated lines represent (a) acceleration, (b) cruising at a constant speed, (c) deceleration and (d) idling (when the engine is running but the vehicle is stationary).

4 EMISSION CONTROL STRATEGIES:
A HISTORICAL PERSPECTIVE

4.1 Engine modification

Vehicle manufacturers were able to meet the first US Federal emission standards simply by regulating the size and tuning of the car engines. The 1970 NO_x standard of 4 g per mile (Figure 8) could be met by modification of the engine, using **exhaust gas recirculation** (EGR). EGR works quite simply by replacing some of the air supply to the cylinder with exhaust gas recirculated into the engine. This will have the effect of lowering the temperature of the reaction (because less O_2 will be present) and hence the amount of NO_x generated (Figure 11). As might be expected, EGR also leads to a reduction in the *performance* of the vehicle.

Engine modifications could not, however, provide the extra control in hydrocarbon and CO emissions demanded by the next phase of the legislation introduced in the 1975 model year (the NO_x standard could still be met by EGR). The solution was provided by the use of oxidation catalysts, which could convert hydrocarbons and CO into CO_2 and H_2O in the exhaust gas *before they* emerged into the atmosphere:

$$2CO(g) + O_2(g) = 2CO_2(g) \tag{17}$$

$$C_8H_{18}(g) + 12\tfrac{1}{2}O_2(g) = 8CO_2(g) + 9H_2O(g) \tag{1}$$

In order to achieve high conversions, a net oxidizing environment was required, so an extra supply of air was injected into the exhaust.

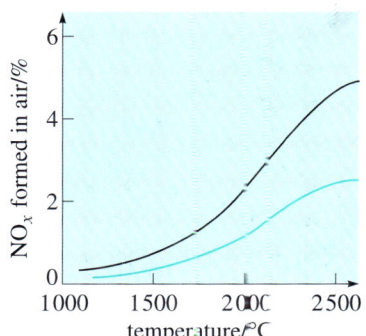

Figure 11   Formation of NO_x from air as a function of temperature.

STUDY COMMENT   Before we get into the bulk of the discussion about the use of catalytic converters, take time to do the following SAQs; they invite you to consider some of the requirements of a catalyst to treat exhaust emissions.

SAQ 3 (revision)   We all have experience of motor cars, in one way or another. See if you can suggest what *practical* properties a suitable catalyst should have for the very challenging application of emission control. (It is not necessary to consider the detailed chemistry involved.)

SAQ 4 (revision)   Using the concepts covered in Block 5, suggest possible candidates for an oxidation catalyst for reactions 1 and 17.

4.2   Oxidation catalysts

In practice, a typical **oxidation catalyst** consisted of platinum and palladium, supported on fine particles of high-surface-area alumina, which in turn were coated onto porous alumina pellets, or onto a **monolith** – an extruded ceramic honeycomb (see Section 4.4.1). Fitting the catalyst was a relatively simple process; it was packaged in a metal canister that was welded into the exhaust system.

Precious metal catalysts were used (in preference to the much cheaper base-metal oxide alternatives) because they were shown to be exceptionally tolerant of most poisons present in vehicle exhausts (such as sulfur), allowing them to meet specifications requiring a catalyst life of 50 000 miles. But there *was* one problem: lead. Lead, added to petrol to prevent knocking (see Box 4), contaminated the catalyst. Thus, the presence of a catalyst required the use of unleaded fuel (see Section 7.2.1). This, together with deep public concern about the health risks of lead, has been a major factor in the phasing-out of lead in petrol. The American Clean Air Act, which called for new cars to be fitted with converters, stipulated that all major gasoline outlets would have to sell at least one grade of unleaded fuel.

In the UK, there has been a substantial progressive switch to lead-free petrol in recent years. In addition, the permitted lead content of leaded petrol was reduced in January 1986 from 0.40 to $0.15 \, \text{g} \, \text{l}^{-1}$. As a result, lead emissions have been reduced substantially, from an estimated 7.2 kT in 1984 to 2.2 kT in 1990.

BOX 4
ENGINE KNOCK

Knocking is the phenomenon experienced when there are disorderly explosions in the cylinder, so that the expansion of gases propelling the piston is not smooth. This can damage the pistons and eventually lead to engine failure.

The extent of knocking depends on the fuel used. 2,2,4-Trimethylpentane (TMP or *iso*-octane) is an excellent fuel and causes very little knock. This has been assigned a knock resistance index or *octane number* of 100. Different fuels can be assigned octane numbers by burning them in a test engine. The engine knock under defined conditions is then compared with that of a mixture of TMP and heptane (heptane has an octane number of zero). When the knocking in the TMP–heptane mixture and the fuel under test begin at the same compression ratio, the octane number of the fuel is equal to the percentage of TMP by volume in the TMP–heptane mixture.

The addition of lead, in the form of lead alkyls, provides a cheap and effective way of boosting octane

rating and improving knock resistance. (Lead also acts as a lubricant for the valves.) For example, tetraethyl lead (**1**) gives an octane number of 97, compared with a value of around 90 in the unleaded refined product. When it burns, the alkyl lead forms solid lead oxides. These are deposited on the sparking plugs and cylinder walls, and can shorten the life of the engine. To prevent this, compounds such as 1,2-dibromoethane are added to the petrol. On combustion, volatile $PbBr_2$ is formed, and this is swept out with the exhaust gases.

$$
\begin{array}{c}
CH_3 \\
| \\
CH_2 \\
| \\
CH_3 - CH_2 - Pb - CH_2 - CH_3 \\
| \\
CH_2 \\
| \\
CH_3
\end{array}
$$

1

Following the inroads made into CO and hydrocarbon emissions, US Federal regulations in the early 1980s were aimed mostly at achieving comparable reductions in the levels of emitted NO_x. However, this was found to be impossible using EGR alone, and so more complex emission control systems have been used since 1981 in order to satisfy the stricter standards.

4.3   Dual-bed catalysts

Initially, the required control of NO_x and other emissions was achieved using two catalyst beds arranged sequentially in a single canister. The first bed (again platinum or a mixture of metals) catalysed the conversion of NO_x into N_2 under net reducing conditions, by exploiting the presence of gas-phase reductants (especially CO):

$$2NO(g) + 2CO(g) = N_2(g) + 2CO_2(g) \qquad (18)$$

The second bed contained an oxidation catalyst. An additional supply of air was injected into the exhaust ahead of this second bed to ensure the complete oxidation of the remaining CO and hydrocarbons. However, this two-bed system was made redundant by the development of a single catalyst capable of the simultaneous *three-way* conversion of all three pollutants – the **three-way catalyst**.

4.4   The three-way catalytic converter

4.4.1   Composition

The current three-way catalyst, shown schematically in Figure 12, is generally a multicomponent material, containing the precious metals rhodium, platinum and (to a lesser extent) palladium, ceria (CeO_2), γ-alumina (Al_2O_3), and other metal oxides. It typically consists of a ceramic monolith of cordierite ($2Mg.2Al_2O_3.5SiO_2$) with strong porous walls enclosing an array of parallel channels. A typical monolith has 64 channel openings per cm^2 (400 per in^2). This design allows a high rate of flow of exhaust gases. Cordierite is used because it can withstand the high temperatures in the exhaust, and the high rate of thermal expansion encountered when the engine first starts – typically, the exhaust gas temperature can reach several hundred degrees in less than a minute. Metallic monoliths are also used, particularly for small converters, but these are more expensive.

Figure 12   Schematic diagram of the three-way catalytic converter. The catalytic converter, in a metal canister, is placed in the exhaust system of the vehicle. As the exhaust gases pass through it, they flow through the channels in the ceramic monolith, where they encounter the particles of alumina impregnated with the metal catalysts

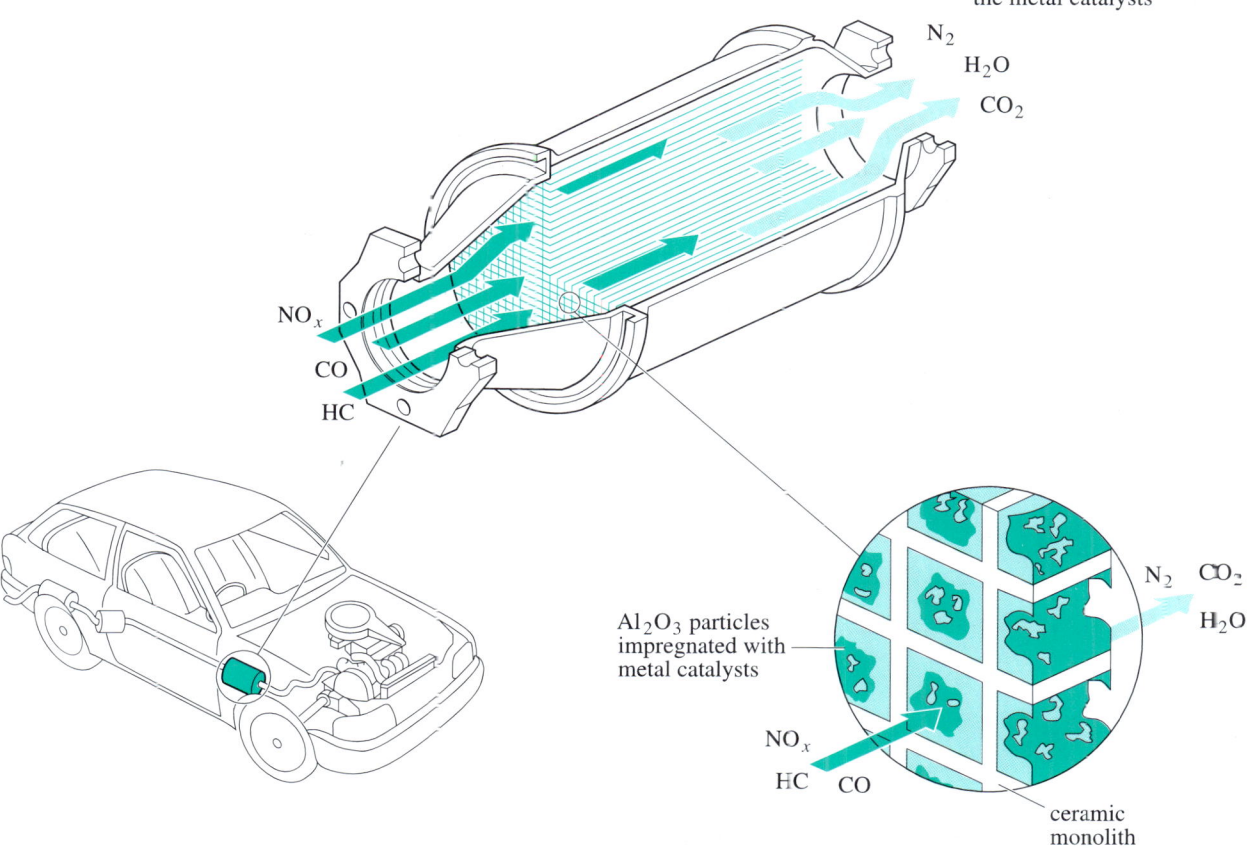

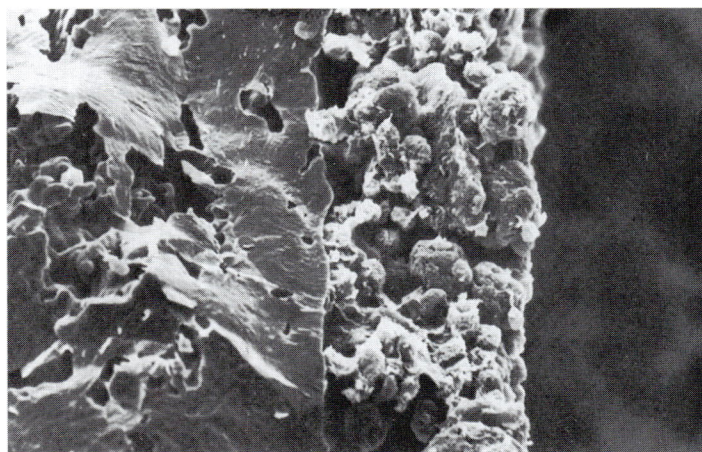

Figure 13 Electron micrograph of a cross section of a ceramic monolith coated with an alumina washcoat.

To achieve a large surface area for catalysis, the internal surfaces of the monolith are covered with a thin coating (30–50 μm) of a highly porous material, known as the **washcoat** (Figure 13). The total surface area is now equivalent to that of about two or three football pitches. The washcoat generally consists of alumina with a large surface area (70–85%), with oxides, such as BaO, added as structural promoters (stabilizers to maintain surface area) and others, for example CeO_2, as chemical promoters (see Block 5). This system becomes the support for the precious metal components (Pt, Pd and Rh). These metals constitute only a small fraction (1–2%) of the total mass of the washcoat, but they are present in a highly dispersed form (see Block 5). They are generally applied by deposition from solution, although they may instead be introduced during formation of the washcoat itself. Exact catalyst formulations are, as one might expect, closely guarded secrets. Some compositions use all three metals; others use Rh together with only one of the other two, typically Pt, as in the present generation of Pt–Rh converters used in the UK, in which Pt constitutes 80–90% of the total precious metal mass.

4.4.2 Catalyst performance

Figure 14 shows the difference in the emission levels for CO, VOC and NO_x for a vehicle, with and without a three-way catalytic converter. It is evident that the catalytic converter reduces the emissions of all three classes of pollutants quite dramatically over a wide range of speeds. Before we discuss the data in any detail, a few words about how they were obtained are in order.

The Federal and European Test Procedures described in Section 3.2 are used to test the emissions from the complete 'finished' converter and engine together, to ensure that a new car model, for instance, will meet the current emissions legislation. Some sort of smaller-scale testing is obviously required in the laboratory. In the research and development of automotive catalysts, activity testing fulfils the function of screening and comparing novel and modified catalysts, and examining their performance under different conditions. The process of screening must provide a reliable means of identifying materials that will perform as active, selective and durable catalysts under automotive conditions. The approach usually taken is to measure conversion of the pollutants as a function of temperature, using a simulated exhaust-gas mixture flowing through a bed of powdered catalyst: the flow-rate has to be high enough to mimic the 'through-put' or *space velocity* of a catalytic converter (typically a *contact time* for the gases with the catalyst of 72 milliseconds is used). The test is then repeated using a different simulated exhaust-gas to represent a different engine mode. Ageing studies are performed by exposing the catalyst to different, and often extreme conditions, for varying lengths of time.

Figure 15 shows a typical graph of catalytic performance over the normal range of operating temperature, 100–600 °C. Until the incoming gases have heated the catalyst to around 250–300 °C, the activity of the catalyst is low. This temperature, at which the efficiency of the catalyst rapidly increases, is known as the **light-off temperature**. Until this temperature is reached, the catalyst is not working at full efficiency, and so CO, NO$_x$ and hydrocarbons will all be emitted from the exhaust pipe in significant amounts. This problem is known as **cold start**. Ideally the light-off temperature should be as low as possible, and in Section 8.1 we shall discuss ways in which manufacturers are trying to realize this objective.

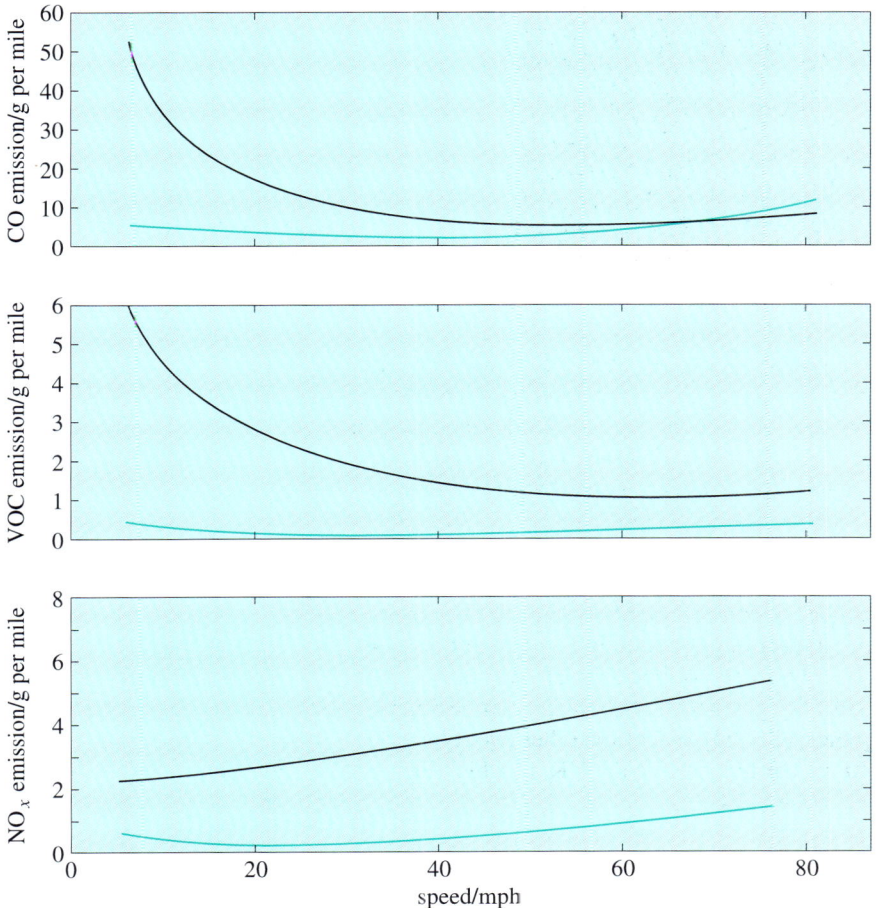

Figure 14   Emission levels for CO, VOC and NO$_x$ for petrol-engined vehicles as a function of speed, with (green) and without (black) a three-way catalytic converter.

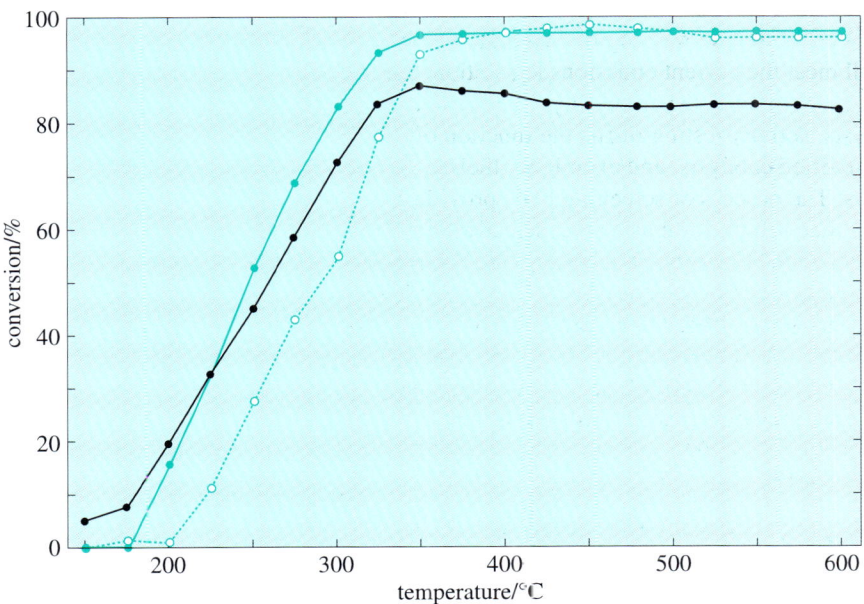

Figure 15   Activity of a three-way catalyst for the simultaneous conversion of CO (black), NO$_x$ (solid green) and the hydrocarbon propene (C$_3$H$_6$) (dotted green).

5 EXHAUST EMISSION CHARACTERISTICS

Before we consider how the three-way catalyst functions in any detail, it is important to understand how the emissions of CO, HC and NO_x *from the engine* depend on the ratio of air (A) to fuel (F) – the **air/fuel ratio** (or **A/F ratio**). The significance of this will become clear when we see that the ratio at which the three-way catalytic converter operates is crucial for its success.

SAQ 5   Taking octane (C_8H_{18}) to be the only constituent of fuel, and assuming that air is 20% O_2 by volume, estimate the stoichiometric A/F ratio (mass ratio) required for total combustion to occur. At this stage neglect the effect of NO as an oxidant. Comment on the difference between the value you obtain and the experimental value of 14.7 : 1. (Use the following relative atomic masses: C, 12.01; H, 1.01; O, 16.00; N, 14.01.)

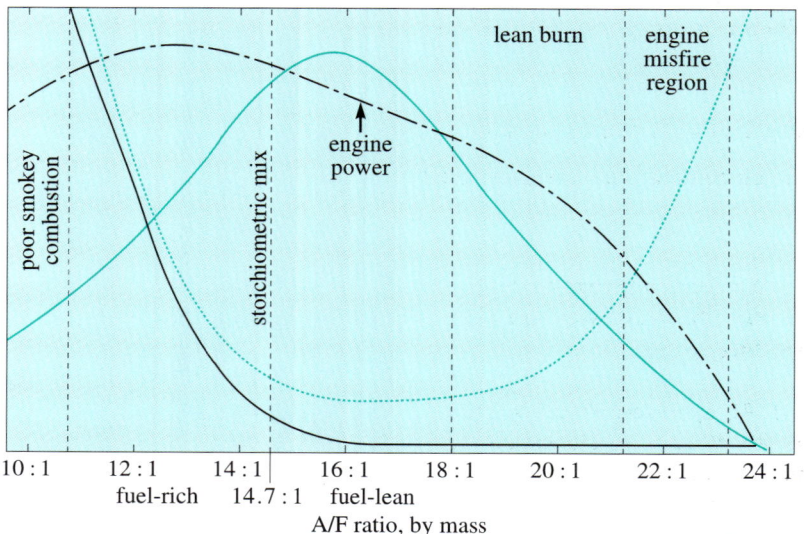

Figure 16   The effect of changing air/fuel ratio on the levels of NO_x (solid green), CO (black) and HC (dotted green) produced *in the engine*. The diagram also shows qualitatively how the engine power output changes with the A/F ratio.

A general relationship between levels of CO, HC and NO_x released from the engine and the A/F ratio is shown in Figure 16. At A/F ratios somewhat *above* **stoichiometric** (14.7 : 1) – that is, when the engine is operating under **fuel-lean**, net oxidizing conditions – low levels of HC and CO are produced in the engine, and there is a peak in NO_x concentration. At higher A/F values, NO_x falls, but the hydrocarbon concentration increases as the engine begins to misfire.

■   Why do you think the CO and HC levels in Figure 16 increase under **fuel-rich** conditions; that is, at low A/F ratios?

▨   The levels released in the engine increase because, below the stoichiometric ratio, there is insufficient oxygen present for total combustion.

When the exhaust gas is close to its stoichiometrically balanced composition, at an A/F ratio of about 14.7 : 1, the concentrations of oxidizing gases (NO and O_2) and reducing gases (HC and CO) are matched; in theory, it should then be possible to achieve complete conversion to produce only CO_2, H_2O and N_2. This is, of course, the objective of the three-way catalytic converter, and so, ideally, it should be operated in a narrow band, or *window*, close to the stoichiometric ratio, within which it will promote *simultaneously* the nearly complete reduction of NO_x to N_2, and the nearly complete oxidation of CO and HC to CO_2 and H_2O. Figure 17 shows the catalyst conversion efficiency for all three classes of pollutants as a function of A/F ratio, with the dotted lines defining the window for conversions of 80% and above.

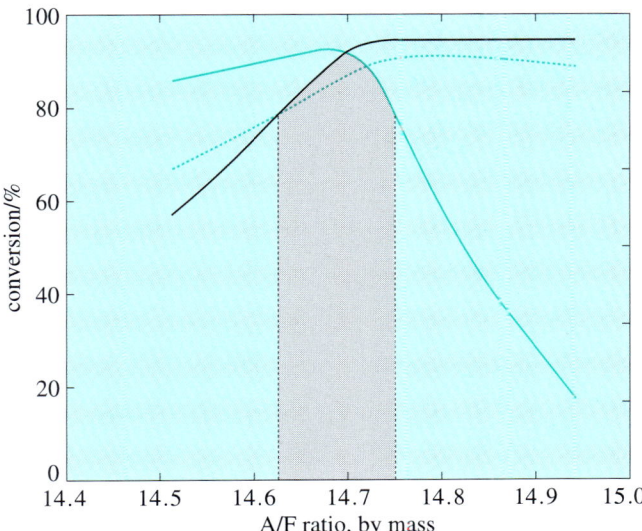

Figure 17   Activity of a three-way catalyst for the simultaneous conversion of NO_x (solid green), CO (black) and HC (dotted green) as a function of the air/fuel ratio. The shaded area defines the window for conversions of 80% and above for all three pollutants. (Note that, for clarity, the A/F ratios are expressed as the amount of air per unit of fuel, e.g. 14.7 instead of 14.7 : 1.We shall use this notation for the rest of this Topic Study.)

■   Ideally, would we want this window to be as wide or as narrow as possible?

▨   A wide window is desirable for catalytic emission control as it lessens the need to tighten the A/F control of the engine.

This window also happens to correspond closely to the optimum range for high performance of the vehicle (see engine power in Figure 16), which was also of growing importance at the time of development (1970s).

STUDY COMMENT   You should approach the following SAQ by thinking about how the mixture expelled from the engine will vary depending on the A/F ratio, and the effect that this will have on the balance of reductants/oxidants present. You should consider how efficient the catalyst will be in converting this mixture, and hence, how this will affect the gases finally emitted from the exhaust (that is, leaving the catalyst)

SAQ 6   Using the information given in Figures 16 and 17, explain the changes in conversion efficiency seen for all three pollutants when the A/F value is (a) greater than the window for optimum conversion, and (b) less than the window for optimum conversion.

Engine control systems have been developed to include an oxygen sensor (or lambda, λ, sensor as it is sometimes called), and an electronic module to regulate the A/F ratio, so that the exhaust composition is kept within the window for optimum conversion. However, because there are time delays in the A/F correction, the ratio cycles very rapidly between slightly fuel-rich and slightly fuel-lean, oscillating about the stoichiometrically balanced composition (14 7 ± 0.3) at a typical frequency of 1 cycle per second. Minimizing the amplitude of the oscillation increases the effectiveness of the converter.

6 THE CHEMICAL REACTIONS

Since its development, the three-way catalyst has been exposed to the full spectrum of techniques available for the characterization of catalytic materials. The data provided can be correlated with the results of activity tests and kinetic measurements, which provide information on the performance of the catalyst. This reveals that although the catalyst functions as a composite material, it can be divided into distinct groups of catalytic centres that provide several different types of site, active for one or more of the many different reactions. The participation of a particular type of site at any given moment will depend on the conditions experienced by the catalyst; for example, whether the gases are a net reducing, stoichiometric, or oxidizing mixture. Measurements of intrinsic kinetics are usually carried out on simple gas mixtures to allow activation energies and reaction orders to be calculated for specific reactions. The data can often contribute to an understanding of the mechanisms by which the surface reactions occur. They are also used to create reaction models that will predict the performance of the catalyst under various anticipated conditions.

The overall reaction scheme is complicated, with many contributing processes. The strategy of the three-way catalyst is to simultaneously remove CO, HC and NO_x, and our treatment will accordingly be divided into three subsections. The desired reactions can be expressed in simple terms as follows:

Removal of CO

CO oxidation:

$$2CO(g) + O_2(g) = 2CO_2(g) \tag{17}$$

Water-gas shift (WGS) reaction:

$$CO(g) + H_2O(g) = CO_2(g) + H_2(g) \tag{19}$$

Removal of hydrocarbons

Hydrocarbon oxidation:

$$C_nH_m(g) + \left(n + \frac{m}{4}\right)O_2(g) = nCO_2(g) + \tfrac{1}{2}mH_2O(g) \tag{20}$$

for example:

$$C_8H_{18}(g) + 12\tfrac{1}{2}O_2(g) = 8CO_2(g) + 9H_2O(g) \tag{1}$$

Steam reforming:

$$C_nH_m(g) + nH_2O(g) = nCO(g) + \left(n + \frac{m}{2}\right)H_2(g) \tag{21}$$

Removal of NO (plus CO or HC (not shown))

CO + NO redox reaction:

$$2NO(g) + 2CO(g) = 2CO_2(g) + N_2(g) \tag{18}$$

or with hydrogen:

$$2NO(g) + 2H_2(g) = N_2(g) + 2H_2O(g) \tag{22}$$

Any number of these reactions may be occurring simultaneously as the A/F ratio goes through its cycle about the stoichiometric composition. The following subsections will look more closely at the removal of each of the pollutants under various conditions, and will also examine the role of the catalyst components.

The supported commercial catalyst is the one most difficult to study because of its complexity, with a large number of different components – Pt, Rh, Al_2O_3, CeO_2, BaO, etc. – present in *one* catalyst. It is therefore often simpler to study *model* systems, such as Pt/Al_2O_3 or Rh/CeO_2, and if certain surface-science techniques are to be used, the 'catalyst' under study has to be even simpler – a particular face of a metal single crystal. These studies, often performed under ultrahigh vacuum (UHV), are far removed from the real catalyst system and the conditions it experiences. Hence, it cannot be assumed automatically that the results will be directly relevant to what is *actually* happening in a converter fitted to an operational vehicle.

6.1 Removal of CO

Under *fuel-lean* conditions (excess O_2), the oxidation of CO has been studied over a very large range of single crystals and model noble metal catalysts, one of the most intensively investigated examples being the Pd(111) surface. Although this metal is not a component of the current three-way catalyst used in the UK, it is worth considering the results in some detail for a number of reasons. The reaction on metals such as Pt is in many ways similar to that on Pd and, in any case, palladium is already being incorporated into future generations of catalytic converter, particularly for the US market. Most notable, however, is the fact that this is one of the few cases in which surface-science techniques have successfully revealed the details of a 'real-world' catalytic mechanism. Specifically, we will see how surface studies of the adsorption of CO and oxygen on Pd(111) – both individually and together – have led to the current understanding of the mechanism of CO oxidation.

LEED results for the adsorption of CO on Pd(111), obtained at room temperature and below, have been interpreted in terms of the structural models shown in Figure 18. One of the significant observations from this work is the readiness with which one arrangement of CO on the surface evolves into another. Thus at a surface fractional coverage of $\theta = \frac{1}{3}$ (Figure 18a), the CO occupies a hollow site where it can bind to three Pd atoms. As θ is increased to $\frac{1}{2}$ (Figure 18b), CO moves out of the hollow to a bridging site, where it binds to two Pd atoms. Finally, at $\theta = \frac{2}{3}$ (Figure 18c), a hexagonal structure forms, in which half of the CO molecules reoccupy hollow sites, while the remainder bind to single Pd atoms at terminal sites. The readiness with which the CO molecules can reposition themselves suggests that the activation energy for surface migration in the chemisorbed state is low, and that CO is a highly mobile species under catalytic conditions.

Oxygen adsorbs dissociatively on Pd(111), and the O atoms are found to be less mobile on the surface than CO molecules. The structure of the chemisorbed layer at maximum coverage is shown in Figure 19.

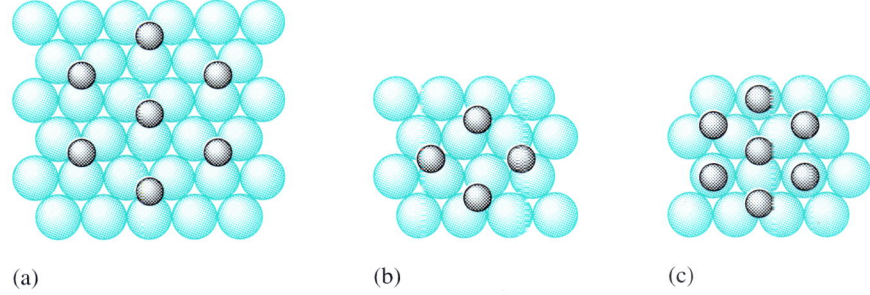

(a)   (b)   (c)

Figure 18   Structural models for the adsorption of CO on Pd(111) at a surface fractional coverage of (a) $\theta = \frac{1}{3}$; (b) $\theta = \frac{1}{2}$; and (c) $\theta = \frac{2}{3}$.

SAQ 7 (revision)   Identify the adsorbate structure shown in Figure 19 in terms of the ($m \times n$) notation introduced in Block 6 (Section 7.2), and determine the fractional surface coverage, θ, of oxygen atoms.

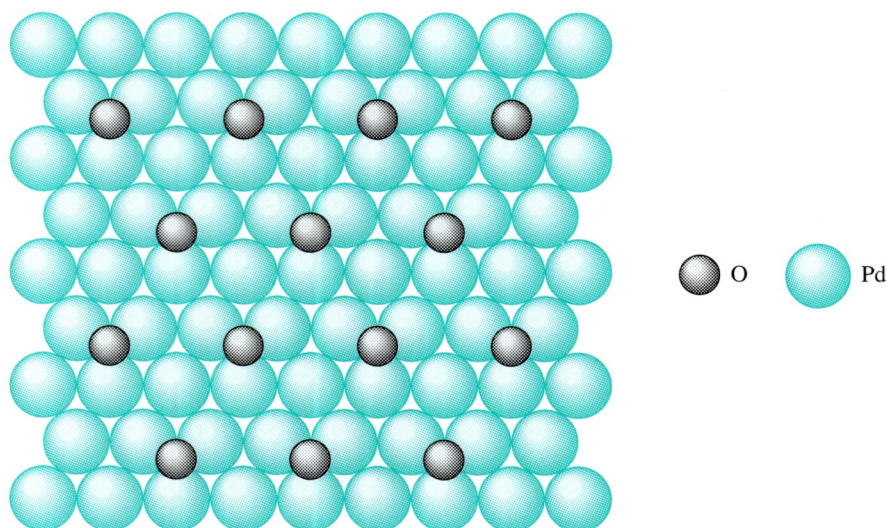

Figure 19   The surface structure of O atoms adsorbed on Pd(111) at maximum surface coverage.

We might now assume that when CO and O_2 are adsorbed together during the oxidation reaction, the properties of the system will be a simple combination of those of the two molecules adsorbed separately. The surface layer would then consist of mobile CO (maximum coverage, $\theta = \frac{2}{3}$) within a fixed lattice of O atoms (maximum coverage, $\theta = \frac{1}{4}$). The fact that this is not the case, as we shall see below, demonstrates an important point. Because of mutual interactions, the behaviour of two (or more) co-adsorbed species very often differs from their behaviour when adsorbed separately.

In the case of CO and O_2, the order in which adsorption is carried out is significant. If CO is adsorbed first, to a coverage greater than one-third of a monolayer ($\theta = \frac{1}{3}$), subsequent oxygen adsorption is completely blocked. With lower coverages of CO, dissociative oxygen adsorption does occur, but the two species form *separate domains* on the surface (Figure 20a). Oxidation will then take place only at the boundaries between domains, and so it will be relatively slow.

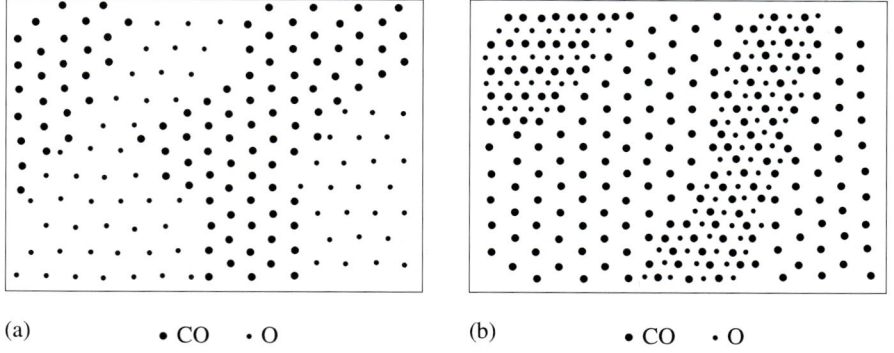

(a)   • CO   · O   (b)   • CO   · O

Figure 20   Schematic representation of domains of CO(ad) and O(ad) on Pd(111).
(a) Separate domains (CO adsorbed first); (b) mixed domains (O_2 adsorbed first).

When oxygen is adsorbed first to its maximum coverage, $\theta = \frac{1}{4}$, subsequent CO adsorption occurs readily and *compresses* the O atoms into domains in which the local coverage reaches $\theta = \frac{1}{3}$. At first, the adsorbed CO is found in separate areas (as when CO is adsorbed first), but as more is added, *mixed* domains form, containing both CO and O, *each at a local coverage of $\theta = \frac{1}{2}$* (Figure 20b).

These mixed domains bring CO(ad) and O(ad) into intimate contact, with the O atoms at *twice* the surface concentration possible in the absence of CO. Thus, the stoichiometry is now that required for the oxidation reaction. Moreover, an electronegative O atom will withdraw charge from the surface. In turn, the surface will withdraw charge from neighbouring Pd—CO bonds, weakening them and so making the CO more readily available for reaction. The net result is that the mixed domain is highly reactive and generates CO_2 at temperatures far below room temperature.

Having thus established that, on Pd(111), rapid CO oxidation *can* occur by way of a Langmuir–Hinshelwood type process (a surface reaction between two adsorbed species), we are almost in a position to propose a detailed mechanism. First, however, we must consider the possibility of an alternative Eley–Rideal type mechanism (Block 5), in which the rate-limiting step involves reaction between an adsorbed species and a molecule in the gas phase. In this case, there are two such possibilities:

$$CO(ad) + \tfrac{1}{2}O_2(g) \longrightarrow CO_2(ad) \tag{23}$$

$$CO(g) + O(ad) \longrightarrow CO_2(ad) \tag{24}$$

■ Experiments show that exposure of a surface saturated with adsorbed CO to gas-phase O_2 does *not* lead to CO_2 formation. How does this simplify consideration of the above reactions?

▨ This observation rules out the Eley–Rideal reaction between CO(ad) and $O_2(g)$, reaction 23.

We can make a decision about reaction 24 on the basis of the information provided in Figure 21. A Pd(111) surface presaturated with a quarter of a monolayer of O atoms was exposed to a beam of gaseous CO, and the surface coverages and the oxidation rate were monitored with time. Figure 21 shows that the rate became significant only *after* a population of CO had built up *on the surface*, and it reached a maximum when the coverages of O and CO were approximately equal. This is clear evidence for a Langmuir–Hinshelwood process. If reaction 24 had been operative, the rate would have been high initially, and would have fallen continuously as the oxygen layer was consumed by the reaction.

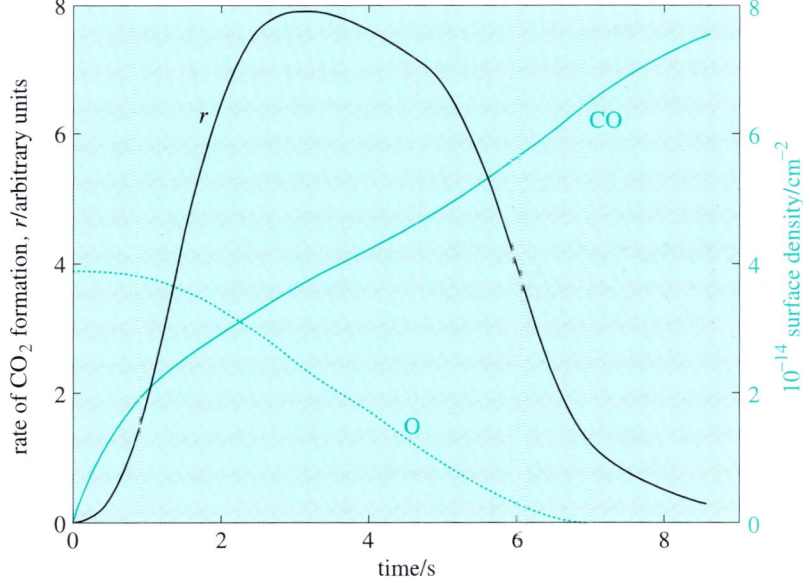

Figure 21 Changes in the rate of CO_2 formation from CO and in the surface density of oxygen and CO on Pd(111). The surface was precovered with a quarter of a monolayer of oxygen atoms at time zero, and then exposed to a constant stream of CO at a pressure of 7.9×10^{-11} atm. (The surface density of an adsorbate on a single crystal surface was defined in Section 7.3 of Block 6.)

Given the research effort that was involved, the mechanism finally proposed for CO oxidation on Pd(111) is deceptively simple:

$$CO(g) \rightleftharpoons CO(ad) \tag{25}$$

$$O_2(g) \rightleftharpoons 2O(ad) \tag{26}$$

$$CO(ad) + O(ad) \longrightarrow CO_2(ad) \tag{27}$$

$$CO_2(ad) \rightleftharpoons CO_2(g) \tag{28}$$

Although the exact nature of the surface intermediates is still not known, the depth of understanding of the catalytic mechanism is quite an accomplishment. But just how relevant are these surface studies to the practical catalysis taking place in the converter?

■ How do the observations outlined below lead to the conclusion that studies of CO oxidation, such as those for Pd(111) above, using surface-science techniques such as LEED, can be of direct relevance to the catalysis taking place in the converter?

1 During the reaction, surface contaminants such as carbon or sulfur are burnt off the Pd catalyst, so that the surface is clean, with the exception of the adsorbed reactants;

2 The catalytic activity does not depend on the crystallographic structure of the Pd metal surface. This is illustrated in Figure 22 where the rate of CO oxidation, over a wide temperature range, is shown to be virtually identical on five different Pd surfaces;

3 The rate of CO oxidation is independent of pressure, depending only on the ratio of the partial pressures of CO and O_2, at least over a restricted range of conditions.

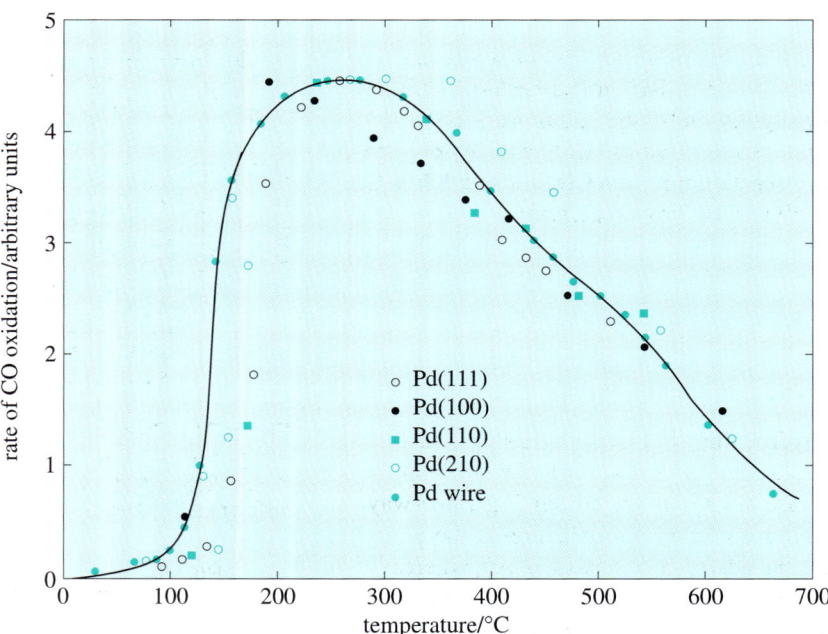

Figure 22 Rates for the catalytic oxidation of CO over a variety of different Pd surfaces.

■ Techniques such as LEED examine adsorption at very low pressures onto *clean* single crystal surfaces, but the real catalytic system is more complicated than this. We are not dealing with a single crystal or a defined crystal plane, and the pressures involved are much higher. However, according to the factors listed above, CO oxidation has been shown to be independent of the Pd surface exposed, and independent of the total pressure of the reactants. In addition, the first observation suggests that under real conditions, the catalyst surface is clean; hence the effects of the adsorbed CO and O are likely to be the only relevant factors. Thus, the conclusions drawn about mechanism from surface studies may not be too far removed from those that apply in the case of the 'real' catalyst.

Indeed, Figure 23 shows that in the case of rhodium there is excellent agreement between the rates of CO oxidation over a Rh(111) single crystal surface and over a Rh/Al_2O_3 catalyst.

Although the sequence of elementary steps is quite simple, the overall kinetics of the CO oxidation reaction is not. The non-uniformity of the surface, and the segregation of the reactants in surface domains, complicates the detailed modelling of the kinetics. The exception is the special case of low surface coverages of CO and O atoms, when they are found to be *randomly* distributed over the surface and so to satisfy one of the criteria for applicability of the Langmuir isotherm. Under these circumstances, Langmuir–Hinshelwood kinetics can be applied.

Figure 24 shows the comparative performance of single-metal catalysts for the oxidation of CO at a fixed temperature. Evidently, all three of the platinum group metals present in automotive catalysts are active for CO oxidation. In addition, results have shown that Rh may improve low-temperature activity. In the current three-way catalyst used in the UK, in which Pt constitutes 80–90% of the noble metal composition and Rh the remainder, it is the Pt that is mainly responsible for CO oxidation.

Under *stoichiometric* or *slightly fuel-rich (reducing)* conditions, where there is insufficient oxygen present to oxidize all of the CO, conversion can also occur by one of the following routes:

- via the CO + NO redox reaction (reaction 18). This will be discussed in detail in Section 6.3.

- via the **water-gas shift reaction** (equation 19), because H_2O is present in the exhaust gases as a product of combustion:

$$CO(g) + H_2O(g) = CO_2(g) + H_2(g) \tag{19}$$

The water-gas shift reaction is catalysed by Pt and/or Rh, with ceria acting as an excellent promoter. $Pt/CeO_2–Al_2O_3$ and $Pt–Rh/CeO_2–Al_2O_3$ are particularly active combinations for the removal of CO under slightly fuel-rich conditions. The hydrogen produced in this reaction will react, in preference to CO, with any oxygen present. Hence, although the water-gas shift reaction removes CO, it also inhibits CO oxidation by producing hydrogen, which will remove any O_2 present:

$$H_2(g) + \tfrac{1}{2}O_2(g) = H_2O(g) \tag{29}$$

6.2   Removal of hydrocarbons

Figure 25 shows a comparative study for hydrocarbon oxidation over single-metal catalysts: it can be seen that Rh, Pd and Pt all give high conversions for A/F ratios at and above stoichiometry. Again (as in the case of CO), in the current UK three-way catalytic converter, Pt is the main component responsible for oxidation of the hydrocarbons. On noble metal surfaces, alkane adsorption is dissociative, whereas unsaturated and aromatic hydrocarbons adsorb either dissociatively or associatively as π-complexes. The subsequent oxidation process is thought to be considerably more complicated than the oxidation of CO, and we shall not consider it in any detail.

When the engine exhaust gas composition is reducing (fuel-rich), hydrocarbons compete effectively with CO for oxygen, and they can also react with water vapour to produce CO and H_2 – a reaction known as **steam reforming**:

$$C_nH_m(g) + nH_2O = nCO(g) + \left(n + \frac{m}{2}\right)H_2(g) \tag{21}$$

This is catalysed by Rh and/or Pt with ceria and, as in the case of the water-gas shift reaction, the combination $Pt–Rh/CeO_2–Al_2O_3$ is particularly active. As we noted earlier, the H_2 produced may react preferentially with any O_2 present, thus reducing the amount of oxygen available to react with hydrocarbons and CO. In addition, the CO produced adds to the burden of carbon monoxide to be removed.

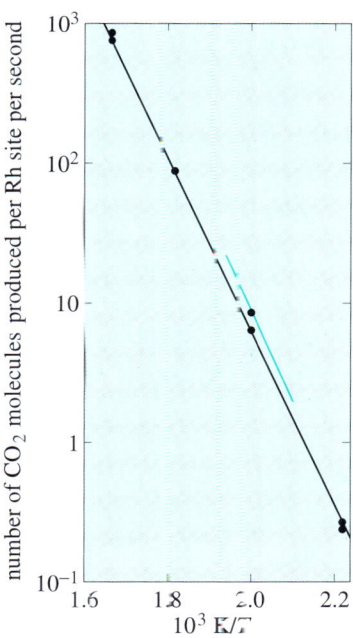

Figure 23   Comparison of the rates for CO oxidation measured over Rh(111) (black) and over 0.01 mass % Rh/Al_2O_3 (green) at $p(CO) = p(O_2) = 0.01$ atm, as a function of temperature.

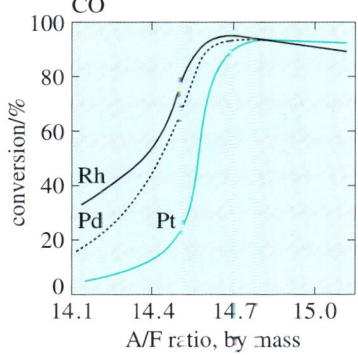

Figure 24   Comparison of catalytic activity for CO oxidation at 400 °C for Pt, Pd and Rh at different A/F ratios.

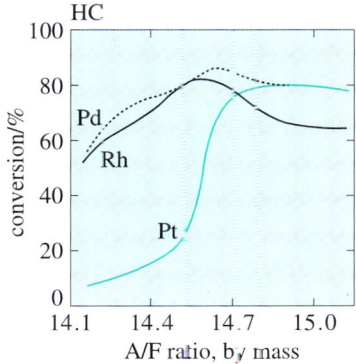

Figure 25   Comparison of catalytic activity for HC oxidation at 400 °C for Pt, Pd and Rh at different A/F ratios.

6.3   Removal of NO

Laboratory experiments have shown that, under the conditions in the catalytic converter, the decomposition of NO to O_2 and N_2 over noble metal catalysts is too slow to be significant. When the A/F ratio is stoichiometric (or below stoichiometry), NO can be removed by reduction with CO and/or hydrocarbons. For simplicity we shall consider only reduction with CO; as with the oxidation reaction, the situation with hydrocarbons is considerably more complicated.

In principle, a variety of products can be formed, specifically:

$$2NO + 2CO = N_2 + 2CO_2 \tag{18}$$

$$2NO + CO = N_2O + CO_2 \tag{30}$$

In addition, H_2 produced from the water-gas shift or steam reforming reactions can reduce NO to N_2, N_2O or NH_3:

$$2NO + 2H_2 = N_2 + 2H_2O \tag{22}$$

$$2NO + H_2 = N_2O + H_2O \tag{31}$$

$$2NO + 5H_2 = 2NH_3 + 2H_2O \tag{32}$$

■   Which reactions should a catalyst *ideally* promote?

▨   The aim of the catalyst is to selectively promote reactions 18 and 22 to produce N_2, rather than N_2O (a greenhouse gas) or NH_3 (a potentially serious general pollutant).

The NO_x activities of Rh, Pt and Pd are shown in Figure 26. It is evident that Rh has the highest activity, particularly under net reducing conditions (low A/F). So why is Rh superior? To answer this question, we need to consider the mechanism of the reaction.

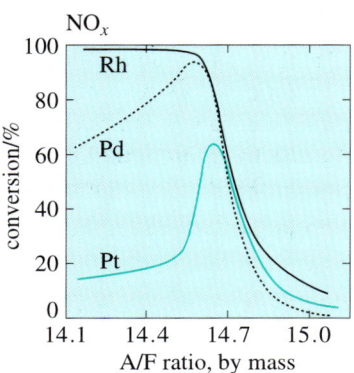

Figure 26   Comparison of catalytic activity for NO_x reduction at 400 °C over Rh, Pt and Pd at different A/F ratios.

The catalytic reduction of NO by CO and/or H_2 over a variety of surfaces has been the subject of a great deal of research. Application of various surface-science techniques has provided some understanding of the elementary steps involved, but the exact mechanism is still controversial. One view is that the first step is the dissociative chemisorption* of NO. The O atoms produced are then removed by the reducing agents CO or H_2. The N atoms can combine to give N_2, react with chemisorbed NO to give N_2O (particularly important at low temperatures), or react with chemisorbed H atoms to form NH_3. These and other processes that may be involved are listed below.

Adsorption

$$NO(g) \rightleftharpoons NO(ad) \tag{33}$$

$$CO(g) \rightleftharpoons CO(ad) \tag{25}$$

Dissociation

$$NO(ad) \longrightarrow N(ad) + O(ad) \tag{34}$$

In the following steps we have assumed, for simplicity, that all products are desorbed as quickly as they are produced. You should recognize, however, that adsorbed species, no matter how transient, will be formed initially.

Surface reactions and desorption

$$CO(ad) + O(ad) \longrightarrow CO_2(g) \tag{35}$$

$$2N(ad) \longrightarrow N_2(g) \tag{36}$$

$$N(ad) + NO(ad) \longrightarrow N_2(g) + O(ad) \tag{37}$$

$$N(ad) + NO(ad) \longrightarrow N_2O(g) \tag{38}$$

* You should note that although we describe this as dissociative chemisorption, strictly it does not meet the definition in Block 5, as NO is in fact first adsorbed associatively and then dissociates *on the surface*.

Reactions with hydrogen

$$H_2(g) \longrightarrow 2H(ad) \tag{39}$$

$$N(ad) + 3H(ad) \xrightarrow{\text{several steps}} NH_3(g) \tag{40}$$

Figure 27 compares the rate of the NO–CO reaction over a Rh(111) single crystal with that over a Rh/Al₂O₃-supported catalyst.

■ After the reaction on Rh(111), the surface was found to have a high coverage of N atoms. What does this suggest about the mechanism?

▨ It suggests that N atom combination (reaction 36) may be the rate-limiting step in the overall NO–CO reaction on Rh(111).

This seems to be the case, particularly at higher temperatures. At lower temperatures, the concentration of NO(ad) increases and reactions 37 and 38 would then be expected to contribute to N atom removal.

The elementary steps 36–38 all require surface mobility of N atoms (to encounter either NO(ad) or other adsorbed N atoms). Although this process may occur on surfaces that are extensive on the atomic scale, such as those of single crystals or large supported crystallites, it has been argued that such mobility will be insignificant on or between the small highly dispersed particles of Rh in the automotive catalyst. Therefore, we might expect the rate-limiting steps and the observed kinetics in the cases of Rh(111) and supported Rh to be different. The Arrhenius-type plots in Figure 27 confirm that this is so: over Rh/Al₂O₃, the reaction has a higher activation energy (the plot in Figure 27 has a larger gradient) and a lower rate (at a given temperature) than over Rh(111).

What then is the rate-limiting step with the supported catalyst? One suggestion is NO dissociation (reaction 34) but there is a more radical alternative, involving a different overall mechanism. Infrared spectra for NO adsorbed on Rh/Al₂O₃ (Figure 28) show bands at 1 743 cm⁻¹ and 1 825 cm⁻¹, which have been taken as evidence of a *dinitrosyl* species, O=N−N=O, formed by reaction 41:

$$NO(ad) + NO(g) \longrightarrow (O{=}N{-}N{=}O)(ad) \tag{41}$$

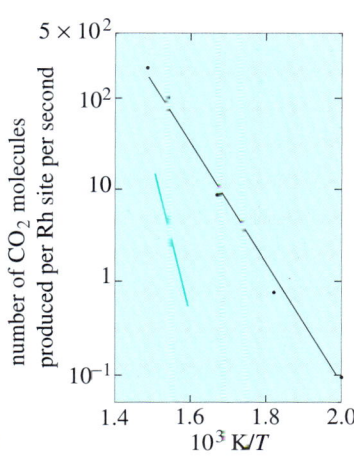

Figure 27   Comparison of the rates for the NO–CO reaction measured over Rh(111) (black) and over 0.01 mass % Rh/Al₂O₃ (green) at $p(CO) = p(NO) = 0.01$ atm, as a function of temperature.

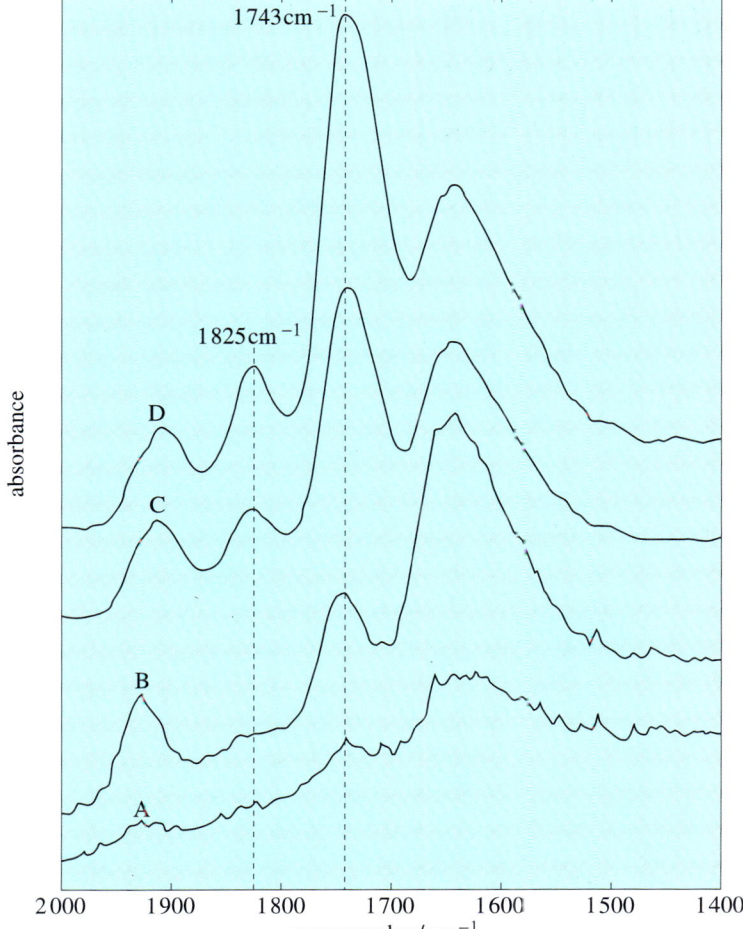

Figure 28   Infrared spectra, recorded at 300 K, for NO adsorbed on Rh/Al₂O₃ as a function of NO coverage, increasing from spectrum A to spectrum D. The bands at 1 743 cm⁻¹ and 1 825 cm⁻¹ have been assigned to the dinitrosyl species O=N−N=O.

This step provides a means, other than diffusion of N(ad), of accomplishing the most important task in the reduction of NO, namely the pairing of two nitrogen atoms on the surface. Once formed, the dinitrosyl species is thought to lose its two oxygen atoms by way of an N_2O intermediate; for example:

$$(NO)_2(ad) \longrightarrow N_2O(ad) + O(ad) \tag{42}$$

$$N_2O(ad) \longrightarrow N_2(ad) + O(ad) \tag{43}$$

To summarize Whichever mechanism is correct – NO pairing to form a dinitrosyl species, or NO dissociation followed by N-atom combination and N_2 desorption – both require catalytic sites that can not only bind NO but also donate charge to the adsorbate. In the first case, this charge would be used to coordinate the two NO molecules. In the second case, it would be transferred into the partially vacant $2\pi^*$ *antibonding* orbital of NO (Figure 29), weakening the N—O bond and hence facilitating dissociation.

On examining the electronic structures of the noble metals, that of rhodium is found to be particularly suitable for facilitating charge transfer to adsorbed NO, with the uppermost occupied electron levels of the metal at higher energy than the partially vacant $2\pi^*$ antibonding orbital of NO (Figure 29). For Pt (and also Pd), however, the situation is reversed. Vacant metal levels lie at energies *below* that of NO $2\pi^*$, so charge will drain *from* this orbital *to* the surface, strengthening the N—O bond.

This picture provides an appealingly simple explanation for the observation that rhodium is the most active of the noble metals for NO_x removal, particularly under net reducing conditions (Figure 26). In fact, not only do Pt and Pd show lower activities, they are also less selective, with Pt, for example, promoting ammonia formation (reaction 32). Rhodium is therefore the essential ingredient in the automotive catalyst for NO_x control. Because it is so active, the amount required is small – about 0.1 mass % of the catalyst, or 0.1–0.2 g, highly dispersed over the surface – but even so, catalytic converters account for around 90% of world rhodium demand (see Box 5).

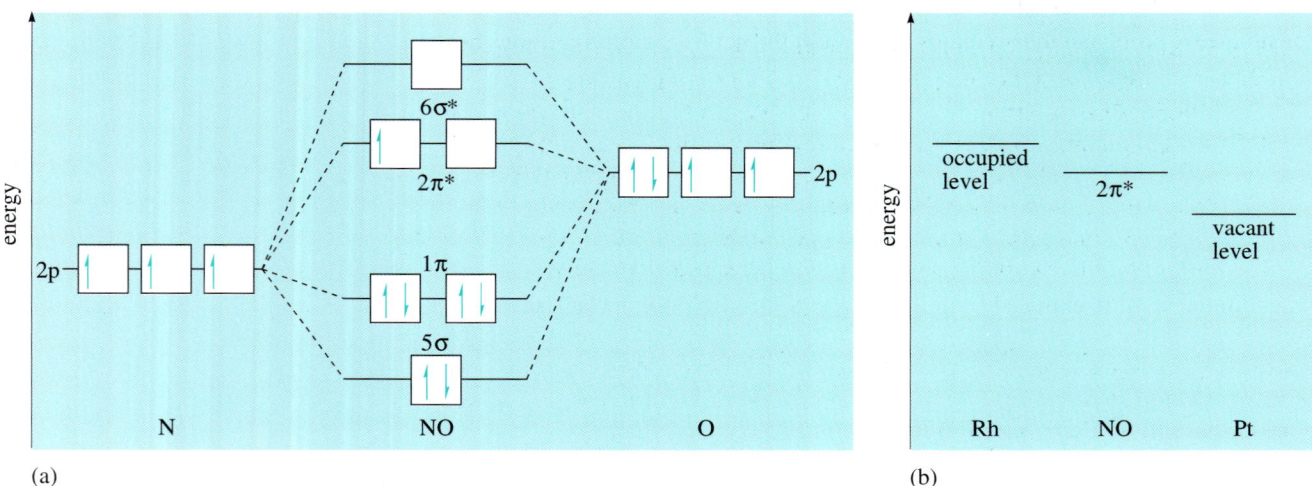

(a)   (b)

Figure 29 (a) The partial orbital energy-level diagram of molecular NO. (b) Schematic energy-level diagram showing the highest occupied level of Rh and the lowest vacant level of Pt, in relation to the $2\pi^*$ molecular orbital of NO.

BOX 5

SUPPLY AND DEMAND OF PRECIOUS METALS

Pt, Pd and Rh form part of the platinum group metals, along with Ru, Os and Ir; they are closely related in terms of chemical and physical properties, and occur together in nature. The metals are all quite rare, and are found in only a few comparatively concentrated deposits – principally in South Africa, the former USSR, Canada and the USA. In South Africa, deposits are mined specifically for the platinum group metals, whereas in Canada and the former USSR, they are produced as by-products from nickel and copper mining.

Separation of the metals is a long and difficult process because they, and their compounds, are so chemically similar in nature. This difficulty in separation, together with the costs incurred in refining the metals to the high purity required, makes their production an expensive business. In addition, huge amounts of the ore have to be mined and processed; for example, it takes approximately 350 kg of South African ore to yield just 1 g of platinum.

Rhodium is essentially a by-product of platinum mining, and is by far the most expensive metal in the platinum group, with a price of approximately $18 250 per kg in July 1995 (compared with Pt ($15 300) and Pd ($5 500)). Because the ratio of Rh to Pt used in automotive catalysts is richer than the mine ratio, the shortfall in Rh came, until recently, from Pt mined for other uses, for example, jewellery. Automotive catalysts are a major application for platinum group metals, the percentage of total demand being 87% for Rh and 34% for Pt (1992).

6.4 Postscript

It is important to remember that the reactions discussed in Sections 6.1–6.3, and often studied separately, occur *simultaneously* in the presence of all the other exhaust constituents. This may have an effect on the efficiency of the individual reactions. For example, kinetic experiments have demonstrated that NO has a strong inhibiting effect on the rate of the reaction of CO with O_2. Figure 30 illustrates the point for a Rh/Al_2O_3 catalyst.

■ With reference to Figure 30, compare the temperatures at which each reaction reaches a rate of 1×10^{-6} mol CO (g cat)$^{-1}$ s^{-1} (the dotted line).

■ The green line labelled CO–O_2 in Figure 30 shows the CO conversion rate for the CO–O_2 reaction in the absence of NO. This approaches a value of 1×10^{-6} mol CO (g cat)$^{-1}$ s^{-1} at about 230 °C. When NO is present (the CO–NO–O_2 mixture) the overall CO conversion rate decreases, and approaches a value of 1×10^{-6} mol CO (g cat)$^{-1}$ s^{-1} only at about 290 °C. In fact, this is close to the temperature at which the CO–NO reaction, shown as the dashed line in Figure 30, becomes significant. Therefore, the inhibition of the CO–O_2 reaction is believed to be due to blocking of the active sites by adsorbed NO.

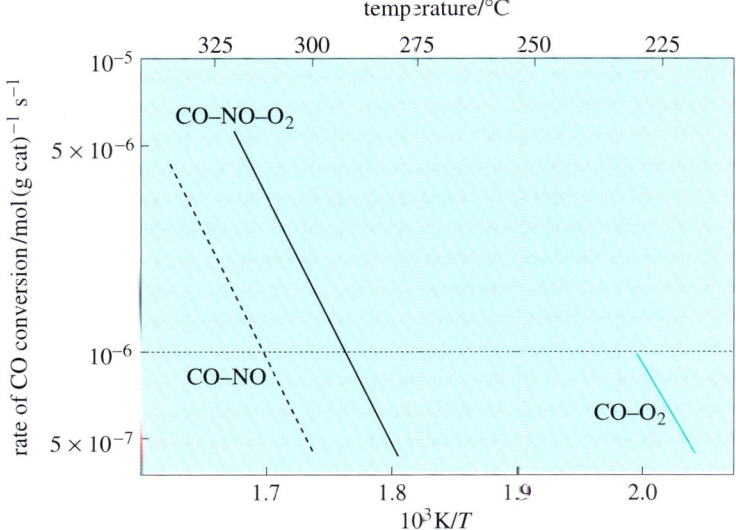

Figure 30   The temperature dependence of CO conversion rates over 0.01 mass % Rh/Al_2O_3 in CO–O_2, CO–NO–O_2 and CO–NO mixtures. The concentration of each of the reactants was 0.5 vol % in all cases.

6.5   The role of CeO$_2$

Figure 31 shows the effect on performance of adding CeO$_2$ to a Pt catalyst for three-way catalytic conversion.

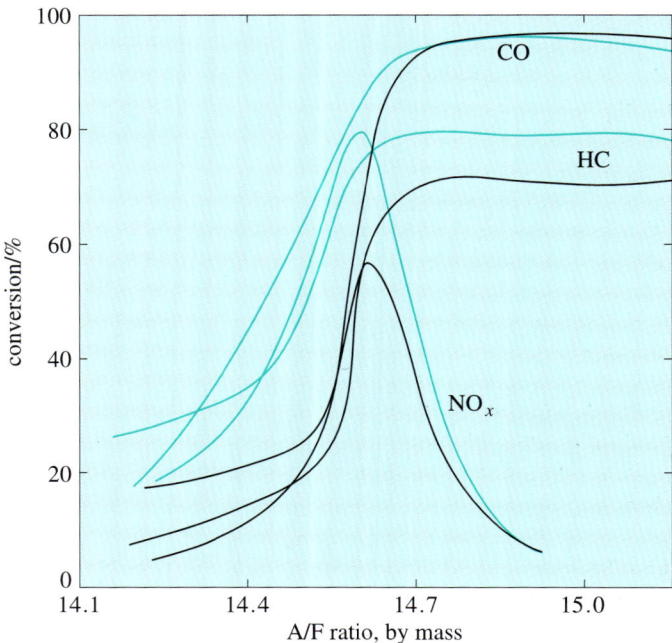

Figure 31   The effects of adding CeO$_2$ to a Pt catalyst in terms of the simultaneous conversion of CO, HC and NO$_x$. The green lines show the conversions for the Pt catalyst with CeO$_2$ and the black lines those without.

■   What is the effect of CeO$_2$ on the conversion efficiencies for CO, hydrocarbons and NO$_x$?

▨   Substantial improvements in all three conversion efficiencies are seen, particularly at A/F ratios just below stoichiometry.

Thus, ceria, which is added with the alumina in the washcoat, is an essential ingredient of the three-way catalyst. It plays a number of roles:

1   Ceria is a *structural promoter,* stabilizing the precious metals and alumina against sintering and particle growth. Figure 32 emphasizes this point.

■   How does the addition of ceria to the catalyst (the coloured line in Figure 32) affect the Pt dispersion?

▨   Clearly, in the absence of ceria, substantial sintering of Pt occurs between 500 °C and 600 °C, causing a sharp reduction in dispersion. Addition of CeO$_2$ results in a significant stabilization of the Pt metal dispersion up to 700–800 °C.

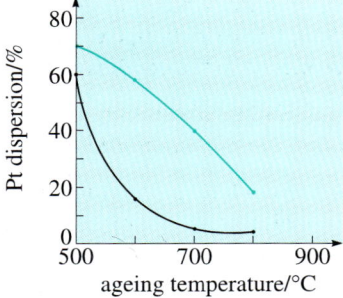

Figure 32   The variation of dispersion stability as a function of ageing temperature, with (green) and without (black) the addition of CeO$_2$ to 0.9 mass % Pt/Al$_2$O$_3$.

Ceria also stabilizes the γ-Al$_2$O$_3$ used in the support, inhibiting a phase change to α-Al$_2$O$_3$, which has a lower surface area. (Lanthanum oxide and/or barium oxide are also often added as stabilizers to help maintain the surface area of γ-Al$_2$O$_3$.)

2   Ceria is known to be able to pick up and store oxygen from the gas phase under fuel-lean operating conditions (excess oxygen) – thus promoting the reduction of NO to N$_2$ – and to release it under fuel-rich conditions (excess fuel), for reaction with CO, H$_2$ or hydrocarbons. Thus, it effectively *dampens* the variations in the A/F ratio as the exhaust gas mixture cycles about stoichiometry, thereby helping to keep operation within the desired window for optimum conversion over the catalyst.

3   As we have seen in Section 6.1, the ceria also enhances the water-gas shift activity of Pt–Rh three-way catalysts, and hence promotes CO removal via the following reaction under fuel-rich conditions:

$$CO(g) + H_2O(g) = CO_2(g) + H_2(g) \qquad (19)$$

The point is illustrated by the results shown in Table 5 for CO conversion under fuel-rich conditions. Increasing the ceria content of the catalyst in the absence of water has no effect, but when water is present the water-gas shift reaction becomes increasingly important. (Addition of ceria also leads to an improvement in activity for steam reforming.)

Table 5   CO conversion for a 1.08 mass % Pt–Rh[a] catalyst on γ-Al$_2$O$_3$ with 1.5, 4.0 and 8.0 mass % Ce levels, under fuel-rich conditions, with and without water present.

	CO conversion/%	
	With H$_2$O	Without H$_2$O
1.5 mass % Ce	54	49
4.0 mass % Ce	64	49
8.0 mass % Ce	70	49

[a] Pt 0.9 mass % and Rh 0.18 mass %.

4   Enhanced conversions of CO, C$_3$H$_6$ and NO at low temperatures have also been observed for Pt/CeO$_2$ catalysts that have undergone a reducing pretreatment. This is believed to be due to an interaction between Pt and CeO$_2$ induced by the reduction, causing an increase in the number and activity of the active sites.

STUDY COMMENT   The enhancement in low-temperature activity is picked up in detail in the video sequence *A clean get-away!* (band 7 on videocassette 2). This would be an appropriate point to view this sequence.

6.6   Summary of Section 6

1   Surface studies of the adsorption of CO and O$_2$ on single crystals and model catalysts have led to the development of a possible mechanism for the oxidation of CO. Dissociatively adsorbed O atoms undergo a surface reaction with adsorbed CO, to form CO$_2$.

2   Under slightly fuel-rich conditions, where there is insufficient oxygen present for complete oxidation, CO can be removed by the water-gas shift reaction, using water produced in the combustion process in the engine. This is promoted by ceria.

3   Hydrocarbons can be removed by oxidation or by reaction with water (a process known as steam reforming).

4   Both CO and hydrocarbons can be removed by reaction with NO under stoichiometric or fuel-rich conditions. The NO–CO redox reaction is believed to proceed *either* by dissociation of NO(ad) followed by N atom combination, *or* by pairing of NO(ad) to give a dinitrosyl species, followed by dissociation. Whatever the detailed mechanism, Rh is particularly active for this reaction, and as such is currently an essential ingredient of the three-way catalyst.

5   Ceria plays a number of important roles in the three-way catalyst: it is a structural promoter, stabilizing both the noble metals and the support against particle growth and sintering; it is an oxygen-storage component, storing oxygen under fuel-lean conditions, and releasing it under fuel-rich conditions; it is a promoter for the water-gas shift and steam reforming reactions; it can enhance the low-temperature activity of the catalyst after certain types of pretreatment.

STUDY COMMENT   The following SAQs invite you to collect together the different roles proposed for each component of the three-way catalyst. It would be a good plan to try them, and check your answers, before moving on

SAQ 8   What are the major roles proposed for the different components of the Pt–Rh/CeO$_2$–Al$_2$O$_3$ three-way catalyst?

SAQ 9   In Figures 24, 25 and 26, Pd in its fresh state is seen to be superior to Pt for the conversion of all three pollutants. Considering the properties we require of a catalyst (see SAQ 3), what possible reason can you suggest for the current widespread use of Pt in three-way catalysts, rather than Pd?

7 CATALYST DETERIORATION

In the UK, three-way catalysts must currently (1996) meet emission standards for a life of 50 000 miles; however, research efforts and legislation are set to double this requirement in the very near future to the current US standard of 100 000 miles. The catalysts do **deactivate** with use. Indeed, the ability to withstand mild deactivation is built into the design of the catalyst, and into the entire emission control system in the vehicle. This is done by setting up vehicles to operate at efficiencies well above the legal requirements at low mileage, so that, as the catalyst slowly deactivates, it will still meet the emission standards.

However, the catalyst may be exposed to conditions that result in more severe deactivation, above and beyond that which is 'allowed for' in its lifetime. The major causes of deterioration are thermal damage (due to exposure of the catalyst to high temperatures), and poisoning by contaminants in the exhaust (notably phosphorus, lead and sulfur). Research aimed at detecting deterioration, and trying to understand its nature, has included post-mortem examinations of used catalysts, and simulated ageing studies, in which the catalyst is exposed to high temperatures or catalyst poisons.

7.1 Thermal effects

Exhaust catalysts usually operate in the temperature range 150–600 °C, but they can experience temperatures of up to 1 000 °C. The conditions that can produce such high temperatures include repeated misfire (resulting in the oxidation of large amounts of unburned fuel over the catalyst), and high driving speeds. In addition, very high temperatures may be experienced if the catalyst is 'close coupled' to the engine, which is one of the possible solutions to the cold-start problem (see Section 8.1). Although commercial catalysts are designed to withstand occasional high-temperature operation, prolonged and repeated exposure to temperatures in excess of 800 °C, especially under oxidizing conditions, have a number of serious effects.

High temperatures may affect all the components of the catalyst. The noble metal particles may sinter (recall Figure 32 in Section 6.5), resulting in a decrease in the fraction of the metal available for catalytic reactions. Such sintering particularly affects the low-temperature activity of the catalyst. It can be countered to some extent by the addition of ceria as a structural promoter, which also stabilizes the alumina support against sintering. However, the ceria may itself undergo crystallite growth at elevated temperatures. This can be inhibited by the addition of barium and zirconium, as shown by the comparison in Figure 33: it is apparent that barium and zirconium help to stabilize the catalytic activity.

At very high temperatures the support may itself sinter or even undergo a phase change, affecting the total surface area. Mechanical loss of catalyst support material may also occur in cases where the washcoat shrinks and cracks, causing it to separate from the monolith (Figure 34a). Again, the problem can be overcome by incorporating so called 'phase stabilizers' – examples include barium and lanthanum – into the washcoat (Figure 34b). At excessively high temperatures the ceramic monolith may even melt, forming additional channels that may allow the exhaust gases to pass through the converter without contacting the catalyst.

High temperatures can also promote damaging interactions between the noble metals, resulting in the formation of a less active alloy, or between a noble metal and base metals in the washcoat support. In particular, it has been established that Rh begins to penetrate the surface of γ-Al_2O_3 at temperatures greater than 600 °C by a solid state reaction between Rh_2O_3 and γ-Al_2O_3. This subsurface penetration and loss of active Rh can be slowed down if the reactivity of the support is minimized by first supporting the Rh on zirconia, ZrO_2, and then incorporating the resulting powder into the γ-Al_2O_3 washcoat. Figure 35 shows the dramatic effect this can have on the catalytic activity of a Rh/γ-Al_2O_3 catalyst. Unfortunately, the incorporation of

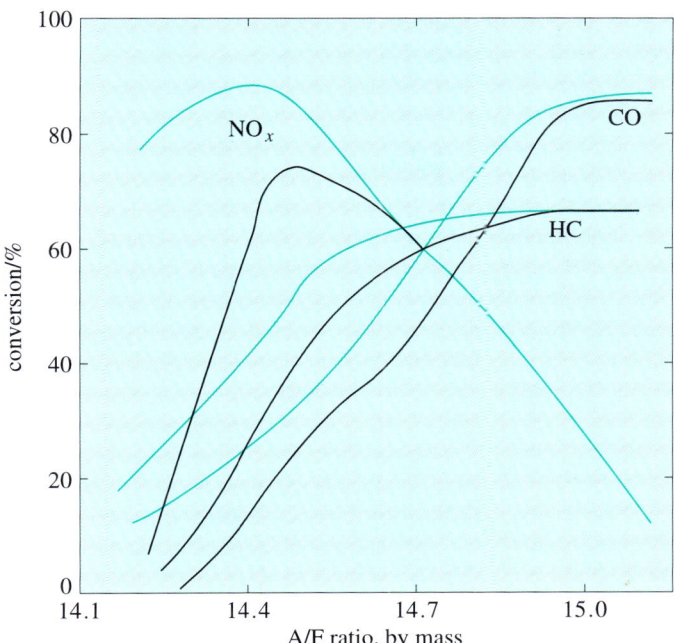

Figure 33   Comparison of catalytic activity for a Pt–Rh/CeO$_2$ catalyst with (green) and without (black) the addition of Ba and Zr, after ageing at A/F = 16.5 at 950 °C for 40 hours.

Rh/ZrO$_2$ into three-way catalysts requires complex manufacturing methods, which are not yet suitable for high-speed production. An alternative approach has been suggested by work that indicates that the Rh/Al$_2$O$_3$ interaction may occur preferentially at the grain boundaries of the support. Ceria can be incorporated as a stabilizer into the alumina in an attempt to preferentially block this interaction.

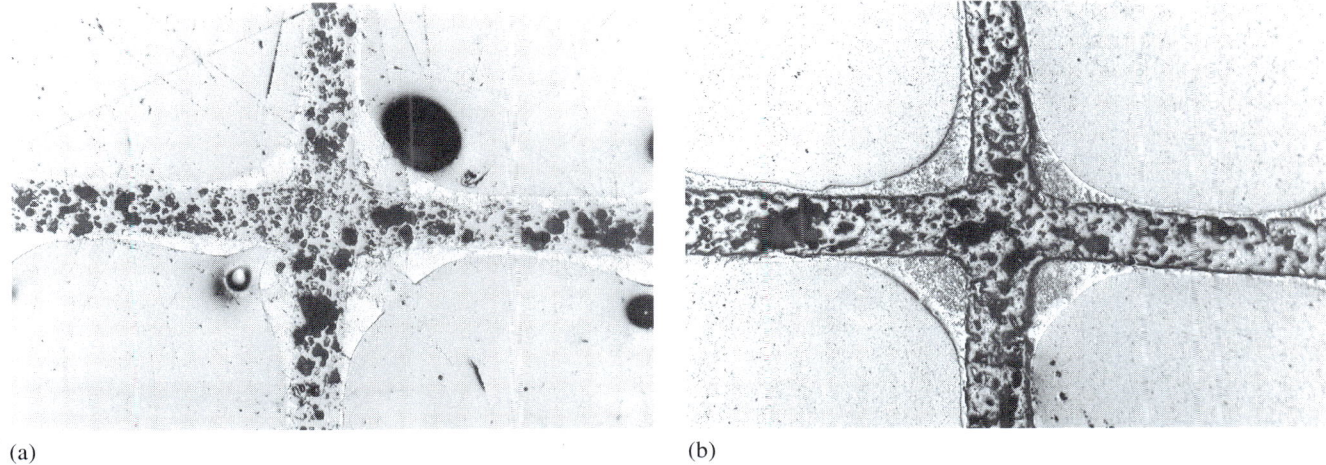

(a)   (b)

Figure 34   Optical micrographs of a washcoat after exposure to high temperatures. In (a) sintering has caused severe shrinkage, whereas in (b) the addition of 'phase stabilizers' has prevented shrinkage cracking.

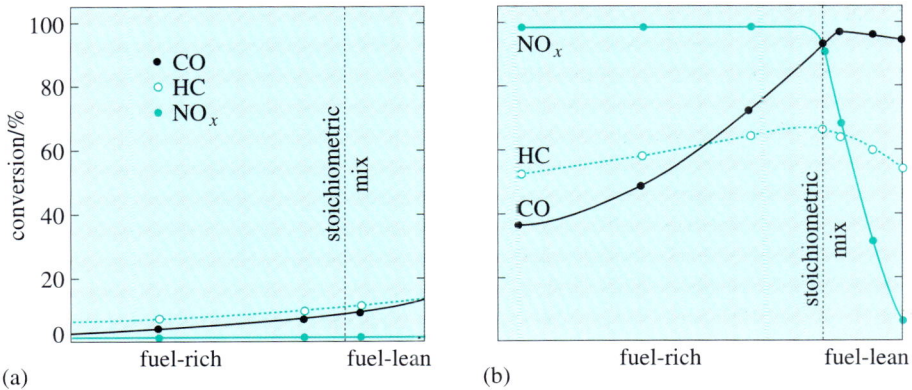

(a)   (b)

Figure 35   Schematic comparison of the catalytic activity of (a) Rh/γ-Al$_2$O$_3$ and (b) [Rh/ZrO$_2$]/γ-Al$_2$O$_3$ after thermal treatment at 1 100 °C in air for 1 h. The activity of the ZrO$_2$-supported catalyst (b) remains virtually intact after the calcination, whereas that of the standard Rh/γ-Al$_2$O$_3$ catalyst (a) is seriously diminished.

7.2 The effect of poisons

7.2.1 Lead

As discussed in Section 4.2, the use of catalytic converters was one of the major contributors to the phasing-in of unleaded petrol. Lead in petrol is a severe poison for the catalyst, and there have been many stories, particularly in the early days of the converter, of people disabling the catalyst by misfuelling. Figure 36 shows how the activity of a typical three-way catalyst is impaired during, and following, intermittent operation with leaded fuel ($0.26 \, \text{g} \, \text{l}^{-1}$) during 15 000 miles of vehicle operation. The efficiency of a control vehicle (unleaded petrol only) was virtually unchanged at 94% for hydrocarbons, 95% for CO and 66% for NO_x. Following the misfuelling, the CO-conversion efficiency (Figure 36a) decreased, but subsequently recovered to an acceptable level. By contrast, the conversion efficiencies for the hydrocarbons (Figure 36b), and especially NO_x (Figure 36c), did not recover to passable values, and hence did not meet the emission regulations current at the time (1986).

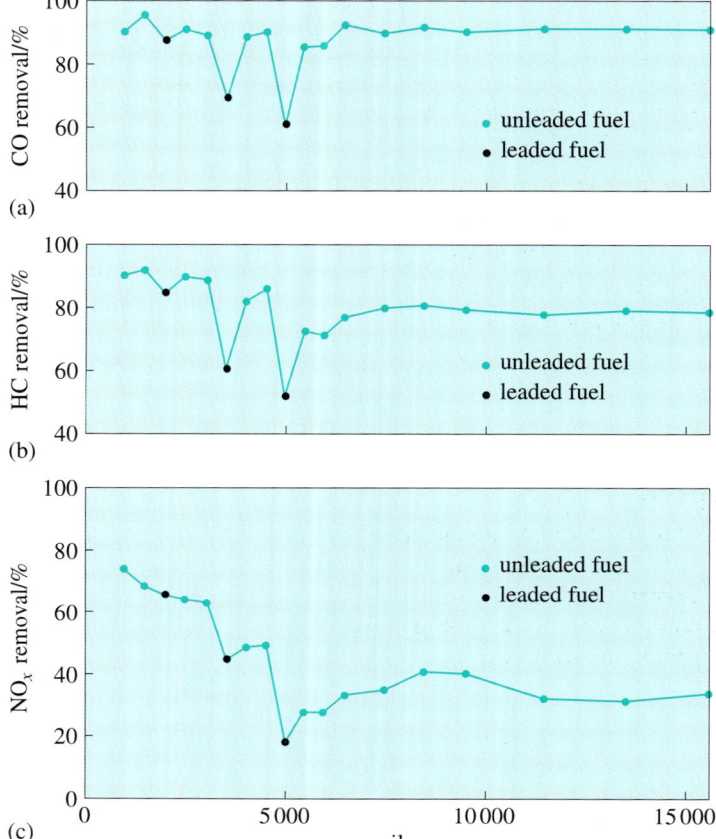

(a)

(b)

(c)

Figure 36 Effect of intermittent use of leaded petrol (black points) on catalyst efficiency (for a typical three-way catalyst) for the conversion of (a) CO (b) HC and (c) NO_x.

Of the various noble metal components, Pd is the most sensitive to lead poisoning. Its activity decreases when there are just trace amounts in the fuel. Rh is slightly less susceptible, and Pt is by far the most resistant. Clues to the mechanism of lead poisoning have come from model systems, which are amenable to detailed surface analysis.

■ Figure 37 shows electron probe elemental maps of $Pt/\gamma\text{-}Al_2O_3$ after exposure to a simulated exhaust gas mixture containing $0.33 \, \text{g} \, \text{l}^{-1}$ of Pb. With which element, Pt or Al, is Pb associated?

□ Pb is associated with Pt, because the Pt and Pb maps are exactly superimposable. Similar results are obtained whether the metal is Pt, Pd or Rh.

This deposition of lead specifically onto the noble metal is believed to occur because the molecules that 'carry' the lead out of the engine, probably halides or oxyhalides, decompose on the noble metal, leaving the lead on the surface.

The fact that Pt is more resistant to lead poisoning than Rh or Pd is largely due to an indirect effect. The small amount of sulfur also present in fuel can act as a scavenger

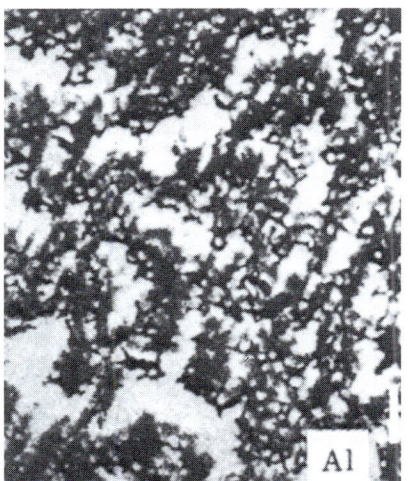 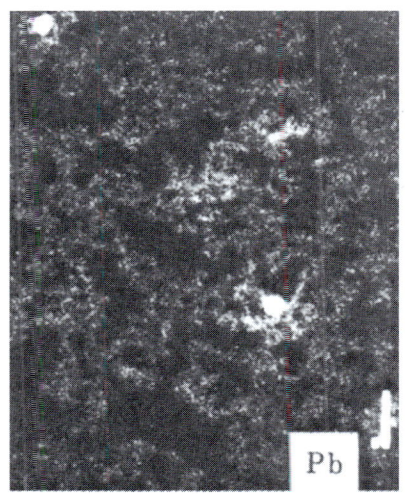 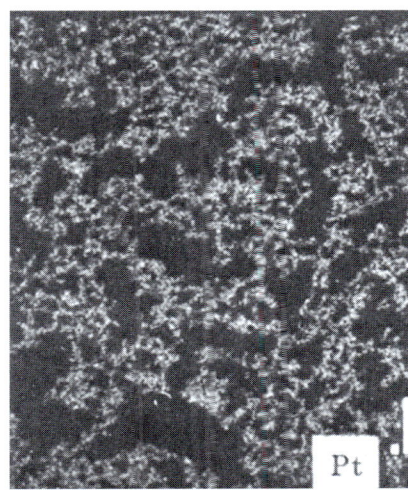

Figure 37 Electron probe elemental maps of Pt/γ-Al₂O₃ after exposure to Pb for 24 hours at 700 °C. Electron probe microanalysis (EPMA) is based on the emission of 'characteristic' X-ray photons following the ionization of atoms by high-energy electrons. EPMA allows quantitative determination of the composition of solids with a resolution of the order of 1 μm. The whiter the area on the map, the higher the concentration of the element under study. The X-ray spectrophotometer is a common attachment on scanning and transmission electron microscopes. (You are not expected to remember the details of this technique.)

for lead. Provided that the sulfur is in its hexavalent oxidation state (S^{VI}), in the form of SO_3, it can combine with lead oxide to form a stable lead sulfate, which, although a poison itself, is not site-specific. Only Pt, however, is a good catalyst for the oxidation of SO_2 (produced from sulfur in the combustion reaction, and present in the exhaust mixture) to SO_3: indeed, it is used for this purpose in the industrial production of sulfuric acid.

7.2.2 Phosphorus

Phosphorus is recognized as a *potential* poison for automotive catalysts. The phosphorus level in fuel is generally very low (2×10^{-5} g l⁻¹), but it is present in higher concentrations in engine oils (1.2 g l⁻¹). Phosphorus derived from the engine oil is believed to react with the alumina support, and also to reduce the activity of the noble metal component. This deactivation is particularly important for Pd, with which phosphorus may form an alloy. At the time of writing, phosphorus levels in engine oil are becoming an issue, and oils with reduced levels are appearing on the market.

7.2.3 Sulfur

Deactivation

The presence of sulfur in the exhaust gas mixture causes a reduction in the activity of the three-way catalyst, particularly for the water-gas shift and steam reforming reactions – processes that are important mechanisms for the removal of CO and hydrocarbons under fuel-rich conditions. Sulfur also decreases the efficiency of NO_x removal. The deleterious effect of exposure to SO_2 on the catalytic activity of a commercial monolithic catalyst (Pt–Rh/CeO₂–Al₂O₃) is evident in Figure 38. Notice, however, that the conversion efficiency recovers quite rapidly once the sulfur has been removed from the gas stream. This is significant because it suggests that a change in the sulfur content of fuel (average 208 mg l⁻¹ in the UK), could achieve a reduction in emissions from vehicles *currently in* use.

Figure 38 The effect of 5 and 20 p.p.m.v. of SO_2 in the simulated exhaust mixture on the catalytic activity of a thermally aged commercial Pt–Rh/CeO₂–Al₂O₃ catalyst as a function of temperature. The test 'after SO_2' is the first test following the removal of SO_2. The hydrocarbon used was propane.

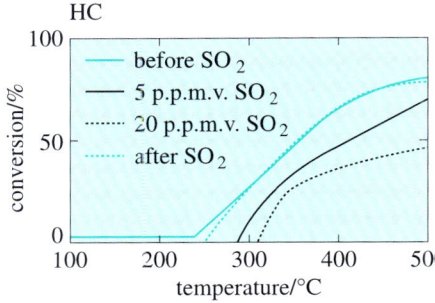 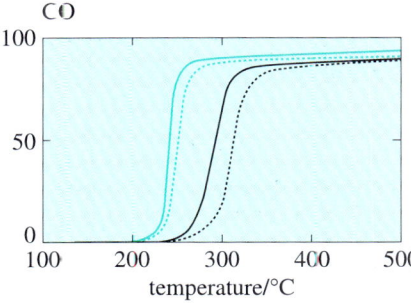 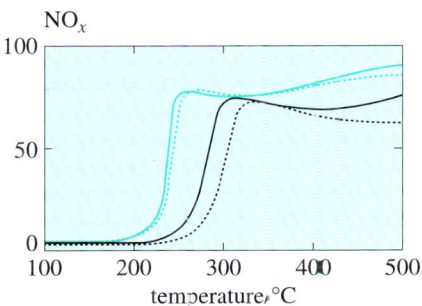

The general resistance of Pt and Rh to sulfur poisoning, and their ability to recover from it, were two of the factors in the original decision in the 1970s to use noble metals rather than the less active, but cheaper, base-metal oxidation catalysts. (In terms of the more stringent legislation now in force, let alone that which will apply in the future, it is unlikely that base-metal activity (or durability) would be sufficient to meet emission control requirements, even if all the sulfur in fuel were to be removed.) Pd is more susceptible to long-term damage, and it is this susceptibility to poisoning that limits its use in the UK at the moment. Pd-only catalysts are, however, under development for the US market – where there are stricter controls over the contaminant levels in fuel.

Some understanding of the mechanism of catalyst deactivation by sulfur has been obtained by examining the effects of SO_2 on the surface area of metals. In Block 5 you saw how the physical adsorption of N_2 can be used to determine the 'total' surface area of a catalyst. As we noted at the time, the capacity of a supported metal catalyst to chemisorb CO can be used as a measure of the free *metal* surface area. It is typically expressed as the ratio *CO/M*, where *CO* is the amount of CO chemisorbed by a fixed mass of catalyst (proportional to the number of *surface* metal atoms), and *M* is the metal content of the catalyst (proportional to the *total* number of metal atoms). The higher the value of *CO/M*, the higher the surface area of the metal. Table 6 gives values of the CO chemisorption capacity of a Pt–Rh/CeO_2–Al_2O_3 model catalyst after ageing in a fuel-rich mixture, in the presence and absence of SO_2.

Table 6 Effect of fuel-rich ageing on the metal-surface area (expressed as *CO/M*) for a Pt–Rh/CeO_2–Al_2O_3 catalyst, in the presence and absence of SO_2. The value of *CO/M* for a fresh sample is 0.79.

Ageing temperature/°C	*CO/M* after ageing	
	In the presence of SO_2	In the absence of SO_2
400	0.20	0.78
550	0.12	0.69
700	0.10	0.39

■ What is the effect of SO_2 on the chemisorption capacity of the catalyst? What do you suggest is happening to the metal surface in the presence of SO_2?

▨ At a given ageing temperature, the CO chemisorption capacity of the catalyst is dramatically decreased in the presence of SO_2. This suggests that some sort of sulfur species has been formed on the metal surface, blocking sites previously available to chemisorb CO, and possibly also inhibiting CO adsorption at neighbouring sites.

Because the effects are not permanent, this sulfur species is presumably relatively weakly adsorbed. As a result it is desorbed, and the activity of the catalyst is regenerated, when the gas–surface equilibrium is shifted in favour of the gas-phase by removal of SO_2 from the gaseous mixture (Figure 38).

The production of hydrogen sulfide

Recently, there has been a great deal of interest in the interactions of sulfur with the three-way catalyst, not so much because of its impact on activity, but rather because of a smelly side-effect. Ever since catalytic converters were introduced, the odour of hydrogen sulfide (H_2S), described as 'smelling like rotten eggs', has been an issue. This is particularly noticeable when a car in front accelerates after idling (at traffic lights or a roundabout, say) or decelerates sharply after cruising. After combustion in the engine, sulfur in the fuel is released to the exhaust gases as SO_2. Under fuel-rich (reducing) conditions, this is converted into H_2S over Pt, the suggested mechanism involving the formation of the metal sulfide as a reaction intermediate. It has been found, however, that the amounts of H_2S generated when engine conditions become fuel-rich, following prolonged running under fuel-lean (or oxidizing) conditions, are larger than expected. The catalyst can apparently 'store' sulfur under lean conditions and release it under rich conditions.

The noble metals do not retain any sulfur under these conditions. However, adsorption studies with both model catalysts and the commercial (fully formulated) catalyst have shown that sulfur storage under lean conditions can occur by interaction of SO_2 with both CeO_2 and Al_2O_3 in the support. Under typical conditions, CeO_2 provides the preferred adsorption sites, and the sulfur storage by the catalyst has been found to increase with increasing Ce content. XPS studies (some results of which are collected in Table 7) have been used to study the nature of the species involved.

Table 7   XPS data recorded for various cerium and sulfur compounds.

Sample	History	Binding energy/eV	
		Ce 3d	S 2p
CeO_2	550 °C, air	881.7	–
CeO_2	20 °C, SO_2	881.6	168.0
CeO_2	550 °C, SO_2	882.8	168.8
$Ce(NO_3)_3.6H_2O$		883.3	–
S^{VI} compounds		–	168.5–168.9
S^{IV} compounds		–	166.3–167.9

■   Comparing the S 2p binding energies with the typical values for S^{VI} and S^{IV} compounds included in Table 7, how would you assign the sulfur species present when CeO_2 is exposed to SO_2 at 20 °C and 550 °C?

□   The typical values suggest that at 550 °C, the sulfur exists as S^{VI}. The value at 20 °C is slightly harder to assign, but it may be attributed to S^{IV} associated with a highly charged cation such as Ce^{IV} or Ce^{III}.

■   Comparing the Ce 3d binding energies with those for CeO_2 in air and for $Ce(NO_3)_3.6H_2O$, what do you conclude about the Ce species present when CeO_2 is exposed to SO_2 at 20 °C and 550 °C?

□   The Ce 3d binding energy of CeO_2 in SO_2 at 20 °C is similar to that of Ce^{IV} oxide (CeO_2 in air). The value at 550 °C agrees reasonably well with that for Ce^{III} in $Ce(NO_3)_3.6H_2O$, suggesting that reduction of some of the surface sites to Ce^{III} may have occurred.

Hence, interpretation of the XPS data suggests that, at room temperature, SO_2 is adsorbed on CeO_2, possibly to form a sulfite species (that is, a species containing S^{IV}). At 550 °C the Ce^{IV} oxide, CeO_2, appears to participate in a redox reaction with SO_2: the sulfur is oxidized to S^{VI} and the cerium is reduced to Ce^{III}, possibly resulting in the formation of cerium(III) sulfate, $Ce_2(SO_4)_3$.

The sulfate/sulfite species formed under fuel-lean conditions with the cerium and aluminium in the support decompose in fuel-rich conditions to release the SO_2/SO_3 species, which are then converted into H_2S over the noble metal component of the catalyst. Catalysts that don't contain noble metals don't produce H_2S.

Although it has been known for some time that automobile exhaust catalysts can produce H_2S in fuel-rich exhaust streams, there has been an increase in the levels emitted in recent years. This is believed to be a consequence of the considerable improvements made in catalyst activity over the years. However, during one study into this effect, it was noted that the H_2S emissions from 'engine-aged' catalysts, that is, those that had been 'on the road' for 50 000 miles, were much lower than those from the fresh catalyst. In addition, the H_2S 'spikes' when going from lean to rich conditions were found to be much smaller on the engine-aged catalyst.

Examination of an aged catalyst revealed traces of the usual poisons, including S and Pb from the fuel, and phosphorus (and zinc) from the engine oil. This suggested that at least one of these components can reduce the storage of sulfur by the Pt–Rh/CeO_2–Al_2O_3 system. Phosphorus appears to be a likely candidate, because it has been found that phosphorus-doped Pt–Rh/CeO_2–Al_2O_3 catalysts exhibit a lower capacity for adsorption of SO_2 than an undoped reference catalyst. This suggests that the phosphorus is somehow 'interfering' with the component of the catalyst that adsorbs the SO_2 – the ceria. Indeed, it has been proposed that a Ce–P–O species (possibly $CePO_4$) is formed, which is more stable than, and hence inhibits the formation of, a Ce–S–O species (for example, $Ce_2(SO_4)_3$). This would have the effect of reducing the storage capabilities of the ceria, and hence of reducing the size of the H_2S spike on going from lean to rich conditions.

To summarize   In view of the discussion above, it would seem that several different strategies could be used to reduce H_2S emissions: (i) decreasing the Ce surface area (however, this will have detrimental effects on the catalyst activity); (ii) improving the A/F control; (iii) reducing the sulfur content of the fuel; (iv) including an H_2S scavenger in the catalyst. (In the USA nickel is used, but in Europe this is prohibited because of concern that it could lead to the formation and release of carcinogenic nickel carbonyl.)

7.3   Summary of Section 7

1   An ability to withstand mild deactivation is built into the design of the catalytic converter. However, severe deactivation could prevent the system from meeting emissions legislation.

2   The major causes of deactivation are thermal damage and poisoning.

3   High temperatures may cause sintering of the metals and/or the support; this can be prevented to some extent by the addition of ceria as a structural promoter. Damaging interactions between the noble metals, or with the support, can also occur at high temperatures. The interaction between Rh_2O_3 and γ-Al_2O_3 can be slowed down by first supporting the Rh_2O_3 on ZrO_2.

4   Lead is a severe poison, particularly for Pd, and is believed to associate with the noble metal.

5   Phosphorus from engine oil can contaminate the catalyst and cause deactivation.

6   Sulfur present in fuel has two major undesirable effects. It can cause deactivation of the catalyst, and it also leads to generation of H_2S.

7   Sulfur is oxidized to SO_2, which is believed to block sites on the metal surface, forming a weakly adsorbed species that is desorbed when the sulfur is removed from the gas stream.

8   Sulfur can also be stored under fuel-lean conditions by Al_2O_3, and especially CeO_2, in the support, and released as H_2S under fuel-rich conditions. The sulfur is believed to be stored as $Ce_2(SO_4)_3$. On going to fuel-rich conditions, this species decomposes, releasing SO_2, which is converted into H_2S over the noble metal. It has been found that adsorption of SO_2, and hence the storage capabilities of a Pt–Rh/CeO_2–Al_2O_3 catalyst, are reduced in the presence of phosphorus. The preferential formation of a Ce–P–O species (possibly $CePO_4$) inhibiting formation of the Ce–S–O species ($Ce_2(SO_4)_3$) has been proposed.

SAQ 10   Refer back to the *CO/M* values listed in the final column of Table 6. How would you explain the variation in these *CO/M* values with increasing ageing temperature?

8 FUTURE TRENDS

8.1 Research

Research challenges in emission control catalysis have not finished with the development of the *current* three-way catalyst. There is still a great need for further work. In the short term, the key areas for improvement relate to the durability and activity of the three-way catalyst: the current life requirement is set to be doubled to 100 000 miles, so the catalyst must be more durable and resistant to deterioration, for example by poisoning. Catalyst activity also needs to be improved to meet ever-more stringent emission regulations. An obvious target here is to improve the catalyst efficiency immediately after the engine is started – by reducing/eliminating the problem of cold start (Section 4.4.2).

The solution to this problem is likely to be a mixture of both catalytic chemistry and engineering. On the catalyst front, much research effort is being aimed at designing a three-way catalyst that would enable 'light-off' at lower temperatures. This area of research is picked up in the video sequence referred to earlier (*A clean get-away!*). However, other solutions are also being investigated, such as (i) the use of electrical heating to raise the temperature of either the catalyst or the exhaust gas during start-up; (ii) heating the catalyst by placing it closer to the engine, known as 'close coupling' (however, this could result in overheating when the car is fully running); (iii) the inclusion of a molecular-sieve trap, which will retain some of the pollutants (predominantly the hydrocarbons) until the catalyst has reached its optimum temperature range – this is a clever bit of chemistry, requiring the desorption temperature of the trap to be closely matched to the light-off temperature of the catalyst; (iv) recycling systems to recirculate the exhaust gas until the pollutants have been converted.

A knock-on effect of decreasing the deterioration of the three-way catalyst, and improving its activity, may be to lower (or even ultimately replace) the noble-metal content of the catalyst. This would help to conserve the limited supply of noble metals, and obviously would have cost implications too.

Because Rh is so expensive, attempts have been made to replace it with other metals, notably Pd. Pd is less active than Rh, and is very susceptible to lead poisoning. However, in the absence of poisons, it is more durable than Pt in high-temperature, oxidizing conditions. The conversion performances of vehicle-aged (25 000 miles) Pd and Pt–Rh catalysts are compared in Figure 39. For CO and hydrocarbon conversion, there is little difference between the activities of the two catalysts. The main disadvantage of the Pd catalyst, relative to the Pt–Rh catalyst, is the decrease in NO_x conversion under fuel-rich (net reducing) conditions. This disadvantage might, however, be overcome in part by tight control of the A/F ratio. The CeO_2 content in such a Pd catalyst is high, and the NO_x reduction activity can be improved by the addition of lanthanum, barium or strontium. In the USA, such Pd-only and Pd–Rh catalysts are becoming increasingly important – made possible by the lower concentrations of poisons in the fuel.

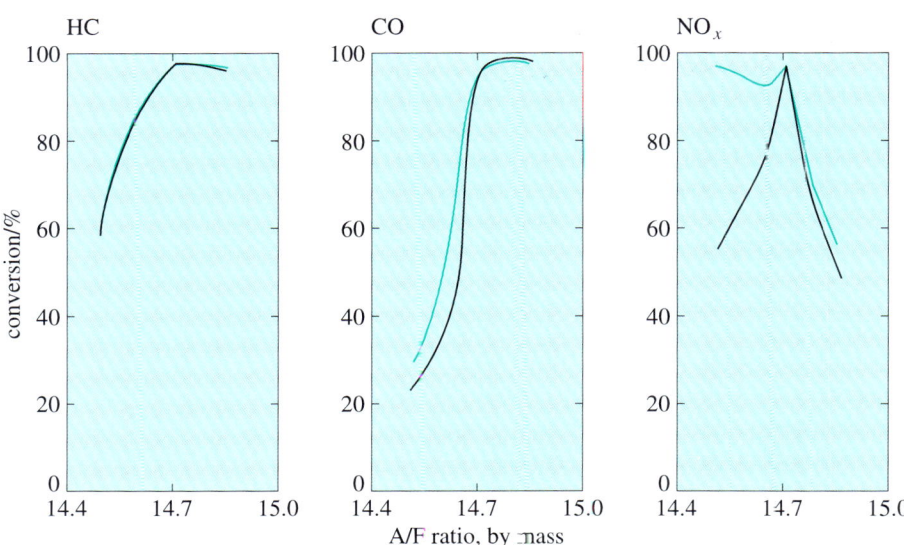

Figure 39   Comparison of the conversion performances of vehicle-aged (25 000 miles) Pd catalyst (black) and Pt–Rh catalyst (green).

Although the three-way catalyst is very effective at reducing the targeted emissions, it represents just one particular stage in the evolving process of the control of atmospheric pollution. Local air quality will improve as the use of the three-way catalyst becomes progressively more widespread, but more effort will be directed at reducing global atmospheric problems (such as the enhanced greenhouse effect). Indeed, many motorists are now demanding greater fuel efficiency, in preference to high performance. Car manufacturers are responding by designing a new generation of *lean-burn* petrol engines – the first models in the UK were marketed in 1994/95. Lean-burn engines are designed to achieve 15% lower fuel consumption, and so to reduce substantially the amount of CO_2 emitted. These engines operate at A/F ratios greater than 17 : 1.

■ Would you expect the current three-way catalyst to be able to operate efficiently at these A/F ratios? You should refer to Figures 16 and 17.

▨ The exhaust gas from these lean-burn engines is expected to contain a high concentration of oxygen, and so the catalytic oxidation of CO and hydrocarbons poses no problem. However, although NO_x emissions from the engine are lower at these A/F ratios than at stoichiometry, there would be insufficient reductants present to react with NO in the presence of so much air.

The challenge now facing catalyst technologists is to design new catalytic converters capable of the reduction of NO_x in the presence of a large excess of oxygen. This is not possible with the current three-way catalyst without going to unacceptable levels of hydrocarbon emissions. One possible solution may be to couple the three-way catalyst with a material capable of converting NO_x into N_2 in ~5% O_2. One such catalyst has already been identified – the copper form of the zeolite ZSM-5. It functions by allowing selective reaction between adsorbed NO_2-type species and retained hydrocarbons. Short-term conversions of NO_x of up to 50% have been obtained. However, Cu/ZSM-5 is not durable under exhaust conditions, particularly in the presence of water; the Cu (which needs to be highly dispersed within the zeolite channels) can agglomerate, and the zeolite structure can undergo hydrothermal collapse. Nevertheless, the prevailing hope is that a detailed understanding of the surface mechanism for the so-called 'lean NO_x' reaction over Cu/ZSM-5 will allow the design of a more stable (non-zeolitic) alternative.

The latest legislation has also seen a reduction in the permitted levels of emissions from diesel engines. In 1994, the US Clean Air Act required a drop in NO_x levels of 20% by 1998, together with a 50% reduction in particulates (both solid and liquid). Diesel engines run in excess oxygen (A/F > 20 : 1), at lower temperatures (150–500 °C), and their emissions are complicated by the presence of particulates and higher SO_2/SO_3 levels. Under these oxidizing conditions, three-way catalysts are inappropriate for NO_x control. Any diesel exhaust treatment must also handle solid, liquid and gaseous components. Diesel oxidation catalysts, currently fitted to fork-lift trucks (since 1967), can oxidize the liquid portion of the particulates, gaseous hydrocarbons and CO, but they do not currently address NO_x removal. This is an important area of current research, but we shall not pursue it further here.

Thus, changing legislation can be seen to be pushing research, as ever-more efficient control systems are required. Recall (Table 4 in Section 3.1) that California already has a programme of future emission standards that requires 10% of all new cars to have zero emissions (NO_x, HC and CO) by 2003.* Thus, the USA has put in motion legislation that is intended to eliminate vehicles as a serious source of air pollution.

8.2   Forecast trends in emission levels

The introduction of catalytic converters in the UK in 1993 has had a major impact on vehicle emissions, with estimates showing that all three classes of pollutant have declined since then (Figure 40). With more vehicles fitted with a three-way catalyst, as new cars replace old cars on the road, and tightening legislation, large decreases are projected for CO and hydrocarbon emissions, at least to 2005.

However, it is evident from Figure 40 that the emissions problem has not been solved.

---

* Technology is not currently available which is economically and technically satisfactory; however, it is likely that electric cars will be required.

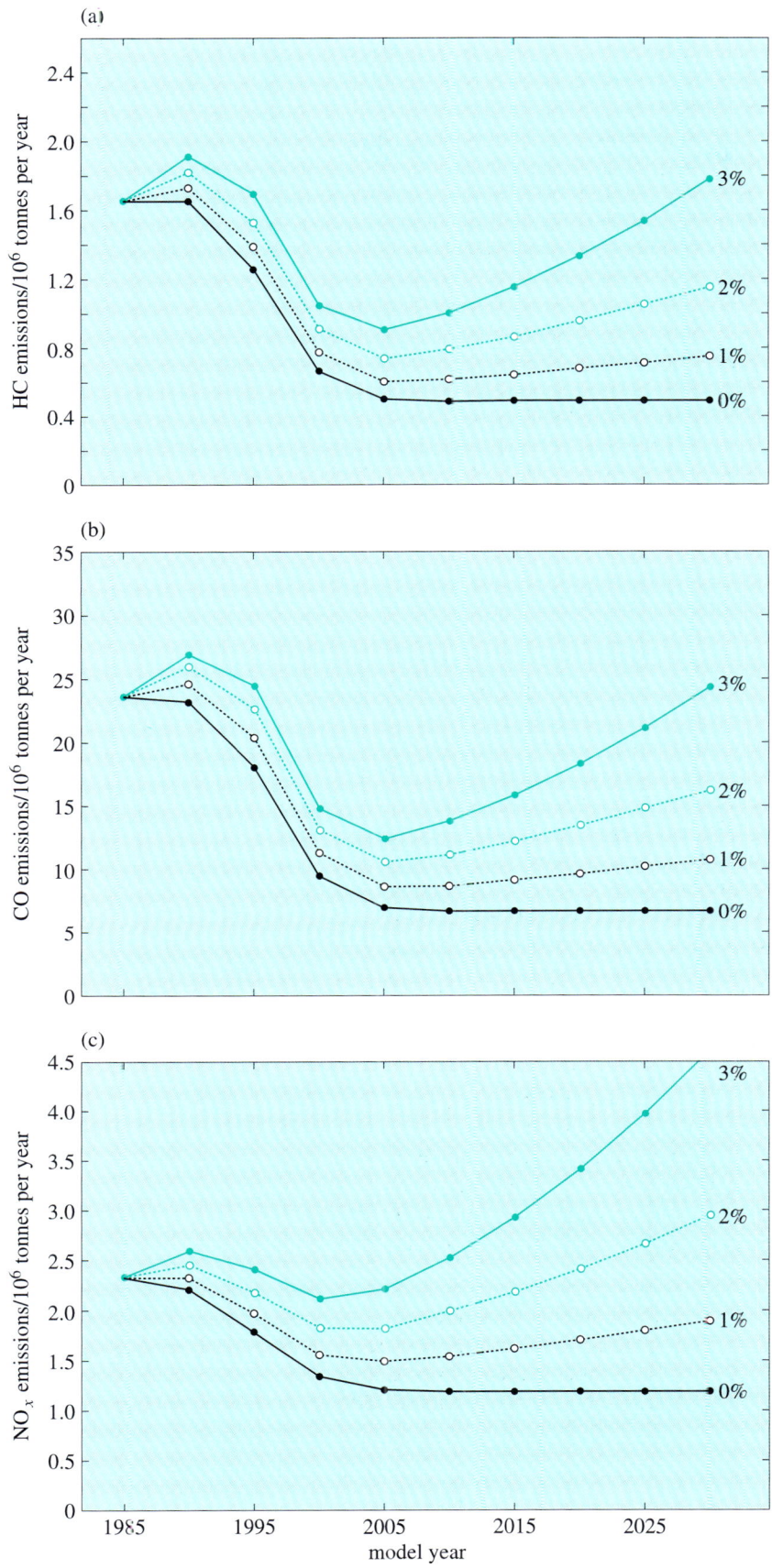

Figure 40   Estimated vehicle exhaust emissions (EU) for (a) HC (b) CO, and (c) NO_x. The figures show trends predicted for zero growth and for annual growth rates of 1%, 2% and 3%.

Vehicle emissions of NO_x, the largest determining factor in future NO_2 concentrations, are predicted to come out of decline in the near future, as traffic growth (there are expected to be about 51 million vehicles in the UK by 2025) offsets the influence of three-way catalytic converters. Similarly, if the current rapid growth of diesel-engined cars continues, particulate emissions, currently forecast to fall owing to stricter limits, are estimated to increase again after 2005. The growth in traffic also makes a substantial contribution to the ongoing accumulation of CO_2 in the atmosphere. Thus, although technology is making huge inroads into the emissions of some of the major pollutants from vehicles, and research will certainly continue, the final solution will almost certainly have to be a political one.

OBJECTIVES FOR TOPIC STUDY 2, PART 1

Now that you have completed Topic Study 2, Part 1, you should be able to do the following things:

1   Recognize valid definitions of, and use in a correct context, the terms, concepts and principles printed in bold type in the text and collected in the following Table.

List of scientific terms, concepts and principles used in Topic Study 2, Part 1.

Term	Page No.
air/fuel (A/F) ratio	22
catalyst ageing	17
catalyst deactivation	36
cold start (problem and solutions)	21, 43
dual-bed catalyst	19
emission control strategies	17
exhaust gas recirculation	17
fuel-lean A/F ratio	22
fuel-rich A/F ratio	22
light-off temperature	20
monolith	18
NO_x	8
oxidation catalyst	18
photochemical smog	9
primary pollutants	6
secondary pollutants	6
steam reforming reaction	29
stoichiometric A/F ratio	22
three-way catalyst	19
washcoat	20
water-gas shift reaction	29

2   Discuss how tropospheric ozone and photochemical smog can arise, and outline the factors that influence their occurrence. (SAQ 1)

3   Outline the ways in which emission control technology has changed with ever-more stringent legislation, and how it is likely to change in the future.

4   Given appropriate information, discuss how the gas mixture expelled from the engine, and the conversion performance of the three-way catalytic converter, depend on the air/fuel (A/F) ratio. (SAQ 6)

5   Apply the principles developed in Block 5 to comment on the requirements for a catalyst to be suitable for emission control. (SAQs 3 and 4)

6   List the chemical reactions whereby the three-way catalyst removes CO, hydrocarbons and NO_x from vehicle exhaust, and outline the roles of the various catalyst components. (SAQs 8 and 9)

7   Apply the principles developed in Blocks 5 and 6 to interpret the results of experimental studies (involving activity tests, kinetic measurements, adsorption studies and/or various surface science techniques) of the three-way catalyst and appropriate model systems. (SAQs 7 and 10)

8   Use the results referred to in Objective 7 to discuss possible mechanisms for the catalytic reactions referred to in Objective 6.

9   Outline the modes of deterioration of the three-way catalyst, and comment on the strategies that could be used to reduce H_2S emissions.

SAQ ANSWERS AND COMMENTS

SAQ 1 (revision and Objective 2)

According to Figure 7, the essential steps in tropospheric CO oxidation can be written as follows:

$$CO + HO\cdot \longrightarrow CO_2 + H\cdot \tag{44}$$

$$H\cdot + O_2 + M \longrightarrow HO_2\cdot + M \tag{45}$$

$$HO_2\cdot + NO \longrightarrow HO\cdot + NO_2 \tag{11}$$

net: $\quad CO + O_2 + NO \longrightarrow CO_2 + NO_2 \tag{46}$

Including the photolysis of NO_2 (reaction 4) and the formation of O_3 (reaction 3):

$$NO_2 + h\nu \longrightarrow NO + O \tag{4}$$

$$O + O_2 + M \longrightarrow O_3 + M \tag{3}$$

the overall effect (obtained by adding equations 46, 4 and 3) can be written:

$$CO + 2O_2 + h\nu \longrightarrow CO_2 + O_3 \tag{13}$$

SAQ 2 (revision)

According to the definition given in Topic Study 1, the mixing ratio by volume (fractional abundance) of NO_2 is given by:

$$\text{mixing ratio} = [NO_2]/[M]$$

where $[NO_2]$ is the concentration (number density) of NO_2. In this case, the mixing ratio is 105 p.p.b.v. $= 105 \times 10^{-9}$, so

$$[NO_2] = (105 \times 10^{-9}) \times (2.5 \times 10^{19}\,\text{cm}^{-3})$$

$$= 2.625 \times 10^{12}\,\text{cm}^{-3}$$

$$= 2.625 \times 10^{12}\,(10^{-2}\,\text{m})^{-3}$$

$$= 2.625 \times 10^{18}\,\text{m}^{-3}$$

The mass of 1 molecule of NO_2 is just the molar mass divided by the Avogadro constant ($6.022 \times 10^{23}\,\text{mol}^{-1}$), so the number density above is equivalent to

$$2.625 \times 10^{18}\,\text{m}^{-3} \times \left(\frac{46.0\ \text{g mol}^{-1}}{6.022\ \times\ 10^{23}\ \text{mol}^{-1}} \right) = 2.00 \times 10^{-4}\,\text{g m}^{-3}$$

$$= 200 \times 10^{-6}\,\text{g m}^{-3} \text{ or } 200\,\mu\text{g m}^{-3} \text{ (as required)}$$

SAQ 3 (Objective 5)

A suitable catalyst should:

(i) operate efficiently over a range of temperatures, including the high temperatures associated with an engine and exhaust system;

(ii) have a long lifetime, ideally equal to the lifetime of the car, but not contribute significantly to the cost of a vehicle;

(iii) achieve desired conversions at high and erratic flow-rates (typically, the entire volume of the exhaust system is replaced 60 000 times every hour);

(iv) resist poisoning and sintering;

(v) be compact, without interfering grossly with the flow of exhaust gases (which would cause power loss from the engine).

It would also make economic and environmental sense to use substances that can be recycled and/or disposed of safely at the end of their useful lifetime.

SAQ 4 (Objective 5)

The platinum group metals (Ru, Rh, Pd, Os, Ir and Pt) would appear to be suitable candidates. Other metals tend to adsorb oxygen so strongly that they are converted into their bulk oxide and rendered inactive. Alternatively, single metal oxides such as V_2O_5, NiO, ZnO and Cr_2O_3 are reported to catalyse the total oxidation of hydrocarbons.

SAQ 5 (Objective 1)

The stoichiometric equation for the complete combustion of octane can be written as follows:

$$C_8H_{18} + 12\tfrac{1}{2}O_2 = 8CO_2 + 9H_2O \tag{1}$$

So combustion of 1 mol of octane will require 12.5 mol of *oxygen*.

Assuming air to be approximately 20% O_2 and 80% N_2 (by volume), the mass of *air* required will be 12.5 $(32.00 + 4 \times 28.02)$ g = 1 801 g.

The mass of 1 mol of octane is $(8 \times 12.01 + 18 \times 1.01)$ g = 114.26 g.

Thus, the A/F mass ratio for complete combustion is:

A/F = 1 801/114.26

    = 15.8:1

This is as close as you would expect to the experimental value of 14.7 : 1, because we have used a very simplified system. We have not included NO as an oxidant or the other hydrocarbons, CO or H_2 as reductants, and we have used octane, *not* the real mix of hydrocarbons in petrol.

SAQ 6 (Objective 4)

(a) Figure 16 shows that over the narrow range of A/F ratios covered in Figure 17 the amounts of CO and HC emitted from the engine decrease as A/F increases. As there is a simultaneous increase in the total amount of oxidants (air + NO_x), the overall conversion of CO and HC increases to approach effective completion at the stoichiometric ratio, and then remains constant in the net oxidizing conditions beyond that point. The sharp fall in NO_x conversion for A/F values approaching and above stoichiometric is understandable in terms of the virtual elimination of reductants in this region. Because the system is unable to remove all of the NO_x, we would expect to see an increase in NO_x emissions from the exhaust. The three-way catalytic converter is therefore unsuitable for engines that run lean, as we shall see again in Section 8.1.

(b) At A/F ratios below the window value there is less NO_x and more HC and CO present in the mixture expelled from the engine (Figure 16). All the NO_x present will react over the catalyst, so the NO_x conversion will still be high (as seen in Figure 17). However, we see a decrease in catalyst efficiency for destroying HC and CO, as there are insufficient oxidants present for complete conversion. We would therefore expect to see an increase in the HC and CO levels emitted from the exhaust.

Obviously, in both cases, in the absence of the three-way catalytic converter the levels of CO, HC and NO_x emitted from the exhaust would be much higher.

SAQ 7 (Objective 7)

The unit meshes of the substrate (1×1) structure and of the adsorbate structure are shown in Figure 41. Evidently, the latter is (2×2) in the $(m \times n)$ notation introduced in Block 6, so the full description of this structure should be Pd(111)(2×2)–O. (Refer to the summary in Section 7.4 of Block 6 – and to Box 2 in particular – if you had difficulty with this question.)

According to the discussion in Block 6 (Section 7.3), *for adsorption on a single crystal surface*, the fractional surface coverage is given by

$$\theta = x / (m \times n)$$

where x is the number of adsorbate species within the $(m \times n)$ adsorbate unit mesh. In this case (Figure 41), O atoms occur only at the corners of the unit mesh, so the latter contains a total of $(4 \times \frac{1}{4}) = 1$ atom. Hence,

$$\theta = 1/(2 \times 2) = \tfrac{1}{4}$$

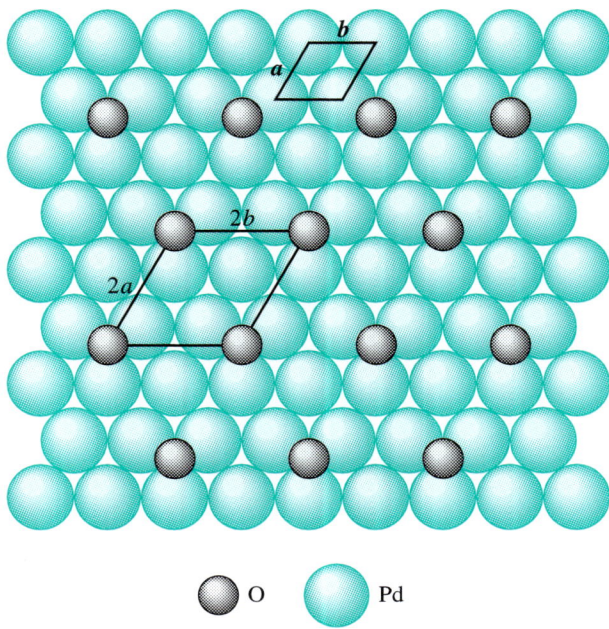

Figure 41   The surface structure of O atoms adsorbed on Pd(111) at maximum surface coverage, showing the substrate unit mesh and the adsorbate unit mesh.

SAQ 8 (Objective 6)

The major roles proposed for the different components of the $Pt–Rh/CeO_2–Al_2O_3$ three-way catalyst can be summarized as follows:

Platinum

- CO oxidation
- HC oxidation
- steam reforming and water-gas shift reactions (with CeO_2)

Rhodium

- CO–NO redox reaction
- steam reforming and water-gas shift reactions (with CeO_2)

Ceria

- oxygen storage
- structural promoter maintaining both metal and support surface areas
- chemical promoter for the water-gas shift and steam reforming reactions
- enhancement in low-temperature activity of Pt after reducing pretreatment (band 7 on videocassette 2)

SAQ 9 (Objective 6)

Pd is in fact cheaper than Pt (Box 5), so the reason for using Pt is not an economic one. The principal properties we require of a catalyst are activity, selectivity and stability. Because we have seen that the first two of these are at least as good for Pd as for Pt, this should lead us to consider the third, stability. In fact, as we shall discover in Section 7.2, Pd has a high susceptibility to poisons, especially lead and sulfur. In addition, it sinters in reducing atmospheres, and can also form an alloy with Rh, reducing the activity of the latter.

SAQ 10 (Objective 7)

The decrease in the value of *CO/M* observed as the ageing temperature increases indicates that the free *metal* surface area of the catalyst is decreasing with increasing temperature. This is likely to be due to sintering of the noble-metal particles, and/or the support.

ACKNOWLEDGEMENTS

Grateful acknowledgement is made to the following sources for permission to reproduce material in Topic Study 2, Part 1:

Figures

Figure 3 Courtesy South Coast Air Quality Management District, California; *Figure 4b* Hanst, P. L. (1975) *A Spectroscopic Study of California Smog*, U.S. Environmental Protection Agency; *Figure 13* Nortier, P. and Soustelle, M. (1987) 'Alumina carriers for automotive pollution control', in Crucq, A. and Frennet, A. (eds) *Catalysis and Automotive Pollution Control*, Elsevier Science Publishers; *Figure 14* Eggleston, H. S. *et al.* (1991) *Corinair Working Group on Emission Factors for Calculating 1990 Emissions from Road Traffic, Volume 1: Methodology and Emission Factors*, Commission of the European Communities; *Figure 15* Courtesy of Dr S. Golunski, Johnson Matthey; *Figure 16* Acres, C. J. K., Thomas, J. M. and Zamaraev, K. I. (1991) *Perspectives in Catalysis*, Blackwell Science Ltd; *Figure 17* Reprinted from *Catalysis and Automotive Pollution Control*, Gandhi, H. S. and Shelef, M. (1987) 'The role of research in the development of new generation automotive catalysts', p. 200, with kind permission of Elsevier Science - NL, Sara Burgerhartstraat 25, 1055 KV, Amsterdam, The Netherlands; *Figures 18 and 20* Reprinted from *Surface Science*, **76**, (2), 1978, Conrad, H. *et al.*, 'Interactions between oxygen and carbon monoxide on a palladium(III) surface'; *Figures 19, 22 and 41* Reprinted from *The Chemical Physics of Solid Surfaces and Heterogeneous Catalysis, Volume 4*, Engel, T. and Ertl, G. 'The role of research in the development of new generation automotive catalysts', 1982, with kind permission of Elsevier Science - NL, Sara Burgerhartstraat 25, 1055 KV, Amsterdam, The Netherlands; *Figure 21* Engel, T. and Ertl, G. 'A molecular beam investigation of the catalytic oxidation of CO on Pd(111)', *Journal of Chemical Physics*, **69**, (3), August 1978, The American Institute of Physics; *Figures 23 and 27* Reprinted from *Journal of Catalysis*, 1986, **100**, Oh, S. H. *et al.* 'Comparative kinetic studies of CO–O_2 and CO–NO reactions over single crystal and supported rhodium catalysts', p. 360, with kind permission of Elsevier Science - NL, Sara Burgerhartstraat 25, 1055 KV, Amsterdam, The Netherlands; *Figures 24, 25, 26, 31 and 33* Reprinted from *Catalysis Today*, 1991, **10**, Funabiki, M. *et al.* 'Auto exhaust catalysts', pp. 34 and 35, with kind permission of Elsevier Science - NL, Sara Burgerhartstraat 25, 1055 KV, Amsterdam, The Netherlands; *Figure 28* Liang, J. *et al.* (1985) 'FT-IR study of nitric oxide chemisorbed on Rh/Al_2O_3', *Journal of Physical Chemistry*, **89**, American Chemical Society; *Figure 30* Reprinted from *Journal of Catalysis*, 1986, **101**, Oh, S. H. and Carpenter, J. E. 'Role of nitric oxide in inhibiting carbon dioxide oxidation over alumina-supported rhodium', p. 114, with kind permission of Elsevier Science - NL, Sara Burgerhartstraat 25, 1055 KV, Amsterdam, The Netherlands; *Figure 32* Reprinted from *Catalysis and Automotive Pollution Control II*, 1991, Diwell, A. F. *et al.* 'The role of ceria in three-way catalysts', p. 142, with kind permission of Elsevier Science - NL, Sara Burgerhartstraat 25, 1055 KV, Amsterdam, The Netherlands; *Figure 34* Courtesy of Johnson Matthey; *Figure 35* Society of Automotive Engineers (1980) *SAE Paper 800843*, Copyright © 1980 Society of Automotive Engineers, Inc.; *Figure 36* McIntyre, B. R. and Faix, L. J. (1986) Lead detection in catalytic emission systems and effects on emissions', *SAE Paper 860488*, Copyright © 1986 Society of Automotive Engineers, Inc.; *Figure 37* Shelef, M. (1987) 'The role of research in the development of new generation automotive catalysts', in Crucq, A. and Frennet, A. (eds) *Catalysis and Automotive Pollution Control*, Elsevier Science Publishers; *Figure 38* Reprinted from *Catalysis and Automotive Pollution Control II*, 1991, Monroe, D. R *et al.*, 'The effect of sulfur on three-way catalysts', p. 612, with kind permission of Elsevier Science - NL, Sara Burgerhartstraat 25, 1055 KV, Amsterdam, The Netherlands; *Figure 39* Society of Automotive Engineers (1989) *SAE Paper 890794*, Copyright © 1989 Society of Automotive Engineers, Inc.; *Figure 40* Walsh, M. P. (1989) 'NO_x emissions from road traffic in Europe', *Workshop on Projections of NO_x Emissions,* Oslo, December 1989; *Figure for Box 2* 'Weather log for 13 December 1991', *Weather*, 1992, **47**, (2), Royal Meteorological Society.

Tables

Table 2 Department of the Environment (1992) *Digest of Environmental Protection and Water Statistics,* No 14, 1991, © Crown Copyright. Reproduced with the permission of the Controller of Her Majesty's Stationery Office; *Table 5* Reprinted from *Catalysis and Automotive Pollution Control II*, 1991, Diwell, A. F. *et al.*, 'The role of ceria in three-way catalysts', p. 145, with kind permission of Elsevier Science - NL, Sara Burgerhartstraat 25, 1055 KV, Amsterdam, The Netherlands; *Table 6* Ansell, G. P. *et al.* (1991) 'Effects of SO_2 on the alkane activity of three-way catalysts', *Catalysis Letters*, **11**, p. 187, © J. C. Baltzer A.G. Scientific Publishing Company; *Table 7* Diwell, F. *et al.* (1987) 'The impact of sulphur storage on emissions from three-way catalysts', *SAE Paper 872163*, Copyright © 1987 Society of Automotive Engineers, Inc.

The S342 Course Team gratefully acknowledge the help of Dr Stan Golunski and Dr Andrew Walker in the preparation of Topic Study 2, Part 1.

Physical chemistry: principles of chemical change

Topic Study 2 Part 2
Dynamic Surfaces

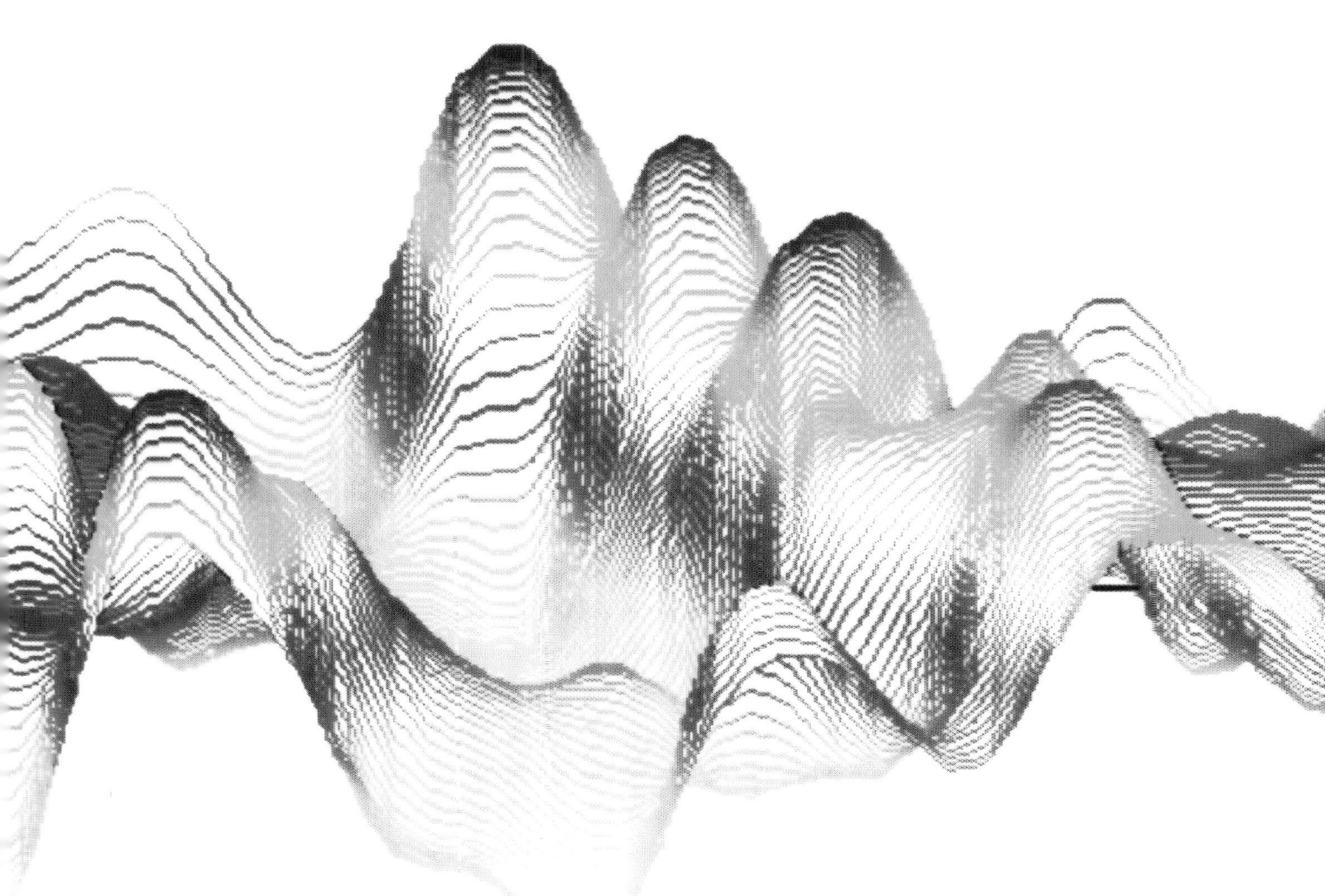

CONTENTS

1 INTRODUCTION

Increasingly, our understanding of heterogeneous catalysis rests on the modern subject of surface science. This new approach, which examines the surfaces of single crystals in great detail has only been made possible over the past thirty years through the development of the ultra high vacuum (UHV) methods described in Block 6. Yet this approach was heralded in 1922 by Irving Langmuir:

> 'In general, we should look upon the surface as consisting of a checkerboard...'

Remarkably this was written before any experimental investigation of a 'plane surface' could be made, and it laid the foundation of chemisorption studies for years to come. The 'checkerboard' model in which adsorption takes place at particular sites is also due to Langmuir:

> 'The surface of a metal contains spaces according to a surface lattice.'
> 'Adsorption films consist of atoms or molecules held to the atoms forming the surface lattice by chemical forces.'

It is on these principles, stated in 1915, that the Langmuir adsorption isotherm (Block 5) is based. Implicit in this type of model are two assumptions that the atoms on the surface of a single crystal face are located at exactly the positions predicted by the structure of the bulk of the crystal; and that the checkerboard is a perfect surface devoid of defects. In spite of the wide applicability of the adsorption isotherm, both assumptions are invalid. All surfaces are to some extent imperfect due to steps, holes or adatoms (Block 6). More importantly for our purposes, no surface is simply an extension of the bulk structure.

- ■ In what respect does an atom on a surface differ from one in the bulk?

- ▣ In the bulk, an atom is surrounded by others on all sides, whereas on the surface there are no atoms above it.

This situation has important consequences for the surface layer because an atom on the surface is bound differently from an atom in the bulk of a solid. This too was recognized by Langmuir, as early as 1916:

> 'The atoms at the surface of a crystal must tend to arrange themselves so that the total energy will be a minimum. In general, this will involve a shifting of the positions of the atoms with respect to one another.'

Fifty years were to pass before this view of a surface could be demonstrated, most dramatically by LEED experiments (Block 6, Section 8). This difference between bulk and surface structure is called **surface restructuring**. We examine various aspects of surface restructuring in Sections 2 and 3.

As an understanding of catalysis is one of the main reasons for studying surfaces, it is important to know how restructuring is related to adsorption. When an adsorbate is chemisorbed on a clean surface, chemical bonds form between the substrate and the adsorbate. Again we may expect the surface layer of the substrate to undergo some reorganization when these new bonds are formed. So restructuring occurs with adsorption. In catalysis, the adsorbed molecules react, either with each other or with molecules in the gas phase, the products being subsequently desorbed. After the removal of the adsorbed molecules, restructuring will again occur to regenerate, at least locally, the original surface structure.

Our understanding of catalysis has undergone enormous change as a result of these ideas. The rigid and static structure of Langmuir's checkerboard has given way to a dynamic model in which surface restructuring occurs due to adsorption and desorption. Surfaces must now be seen as *dynamic* structures, controlled by kinetic processes that are influential in determining the efficacy of the catalyst. In the final Section of this Topic Study, we examine some recent results of dynamic restructuring of relevance to catalysis.

However, before we proceed, it is worth noting two observations relating to the subjects of surface science and catalysis. First, although the study of surfaces, especially of single crystals, has revealed much about existing catalytic processes, no new commercial catalyst has yet been designed as a result of studies in surface science. Secondly, there are few surface techniques that can be used under the conditions at which catalysts operate, especially at high pressure. This so-called pressure gap, between laboratory studies and industrial processes, no doubt has much to do with the fact that, in spite of the recent developments in surface science, catalysis remains an empirical subject.

STUDY COMMENT   This short study is about a topical aspect of research in surface science. The examples used to exemplify it will give you useful practice in applying the principles developed in Sections 7 and 8 of Block 6. There are no associated audio or video sequences.

2 RESTRUCTURING OF CLEAN SURFACES

In the restructuring of a surface, the surface atoms move with respect to the ideal positions defined by the bulk lattice. These displacements may be large or small, and they may be in directions perpendicular or parallel to the surface. Combinations of these movements may also occur. Some terminology distinguishes the extent of movement. Where this is small, such that no bonds are broken, the restructuring is of two types — **interlayer** (or **vertical**) **relaxation**, when the movement is perpendicular to the surface, and **lateral relaxation** (or **displacement**), when it is parallel to the surface. If bonds *are* broken, relatively large movement occurs, and the situation is termed **reconstruction**.

In many cases, the degree of restructuring is dependent on the type of substance and, especially, on the bonds involved. In metals, the bonds are undirected. So, on a metal surface, where the overlying atoms are removed, the surface atoms usually adopt new sites with only small movements from their bulk positions. By contrast, in covalently bound solids the bonds are directional. In this case, in the absence of overlying atoms, a major restructuring usually occurs. We consider examples of both types below.

However, we shall first consider two other reasons for surfaces to have structures that are not the perfect crystal planes that we might predict from our notion of bulk structures. The first of these results from experimental methods and the second from inherent thermodynamic effects.

2.1 Experimental inaccuracy

In surface studies, crystal faces are often produced by cutting or cleaving a crystal in a particular direction in a vacuum in order to expose a clean flat surface of the desired crystal plane. This can never be done with complete accuracy, and often the surface is aligned only within about 0.05° of the correct angle. The result is a stepped surface in which single atom steps occur at intervals, as shown schematically in Figure 1. When the steps occur infrequently, it is not possible to detect them with a LEED experiment.

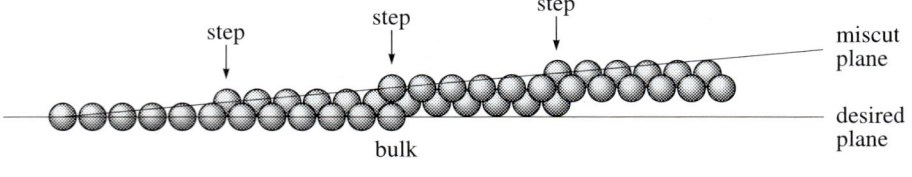

Figure 1 A section through a stepped surface as produced by cleaving at a slightly misaligned angle. The frequency of the steps is much exaggerated; that is, the angle of misalignment of the cleavage is exaggerated.

■ Which technique is best suited for detecting steps of the kind shown in Figure 1?

▪ This problem requires an imaging technique of single-atom resolution. It can be done using scanning tunneling microscopy (STM, Block 6, Section 11).

An example of this approach is shown in Figure 2, the (110) surface of copper in which four single-atom steps occur between left and right.

■ What is the frequency of the single atom steps on this surface? Assume that a copper atom has a radius of 90 pm.

▪ There are roughly four complete terraces across the width of the frame. So each terrace is about 110 nm across. The number of copper atoms in each terrace is therefore $(110 \times 10^3 \text{ pm})/(2 \times 90 \text{ pm}) \approx 600$. As the step height is one atom, the frequency is roughly one in 600.

2.2 Thermodynamic defects

A crystalline solid is an ordered structure, in contrast to a liquid, which has only very short-range order, and a gas which is completely disordered. The breakdown of order, associated with increasing temperature, entails a disruption of chemical bonds, but this is offset by an accompanying increase in entropy. Crystalline solids are often envisaged as 'perfect' structures, but such perfection can only exist in principle at a temperature of absolute zero (0 K). At finite temperatures the perfectly ordered structure breaks down, so that atoms occur where we would not expect to find them in a perfect solid.

As you may recall from the Second Level Inorganic Course, the simplest type of imperfection is the **point defect**, in which an atom leaves its ideal lattice position to occupy a site that is not part of the lattice, a process that can happen in the bulk or on the surface, as illustrated schematically in Figure 3.

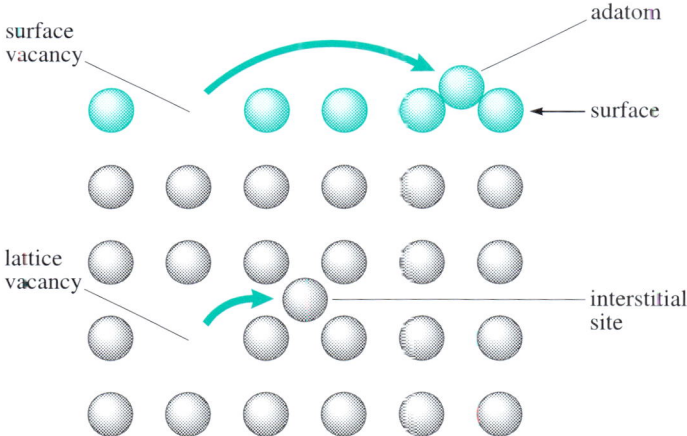

Again, it is STM that enables us to observe this kind of defect. Figure 4a shows a STM image of a Si(100) surface in which there are six steps downward from lower left to upper right, a drop of about 1.4 nm in total. At a magnification ten times greater (Figure 4b), the region around a single step shows kinks in the edge and,

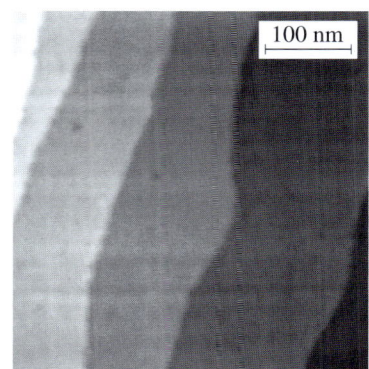

Figure 2 A STM image of the (110) surface of a cleaved copper crystal. The frame measures 440 nm by 440 nm.

Figure 3 A section through a solid, showing how single atom migration within the bulk leads to a lattice vacancy, together with occupation of an interstitial site, and single atom migration from the surface creates a surface vacancy and an adatom.

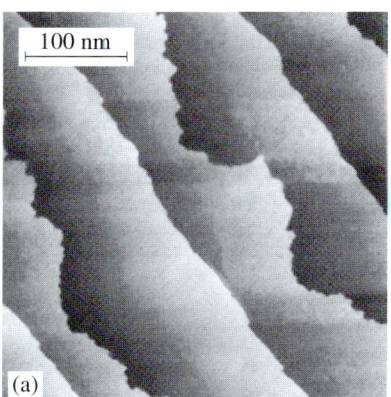

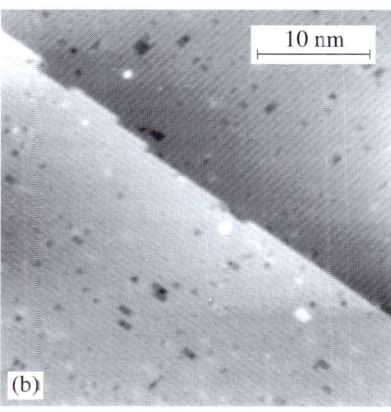

Figure 4 (a) A STM image of a Si(100) surface, showing seven terraces of single atom steps, each 236 pm in height, the diameter of a silicon atom. The frame measures 400 nm × 400 nm, and the mis-alignment of cleavage is about 0.1°; (b) a STM image of one step edge, at ten times the magnification in (a). The frame measures 40 nm × 40 nm.

on the adjoining terraces, a number of point defects. These are both vacancies (dark spots) and adatoms (light spots), the latter possibly resulting from the relocation of the atoms dislodged from the kink and vacancy sites.

■ How would you expect the number of point defects in a crystal to change as the temperature is increased?

▨ For point defect formation, $\Delta H > 0$ (bonds are broken) and $\Delta S > 0$ (more geometric disorder). As $\Delta G = \Delta H - T\Delta S$, ΔG will become more negative with increasing temperature. So the number of point defects increases as the temperature is increased.

For example, in a copper crystal there is about one point defect for every 10^{19} atoms at 300 K, but at 900 K there is one point defect for every 10^6 atoms.

2.3 Interlayer (vertical) relaxation

As indicated above, small changes (relaxation) are observed in metals. It might be expected that the surfaces of all metals would respond to the absence of overlying atoms. But within experimental error, usually ±1%, it is found that the *close-packed* surface layers *fcc*(111) and *bcc*(110) — both shown on the fold-out sheet at the end of Block 6 — have vertical spacings the same as those in the bulk crystal. In this case they are called **bulk-truncated**.

In most other cases, interlayer relaxation is observed in a way that is predictable in bonding terms. Although bonding in solids is described in terms of band theory, the bonding electrons in the surface layers may be regarded as isolated from those in the bulk. This more localized description allows us to make predictions.

■ In the absence of an overlying layer, how would you expect the bonding between layer 1 (the surface) and layer 2 of a metal to compare with the bonding between bulk layers? How would this be likely to affect the spacing of layers 1 and 2?

▨ The absence of any overlying layer leaves the atoms of the surface layer with the electrons that would have been employed in bonding to the overlying layer. These electrons now become available to participate in bonding between layers 1 and 2. The bonds between layers 1 and 2 therefore become stronger, and a contraction between these layers is then expected.

Such a contraction is an example of interlayer (or vertical) relaxation, and is observed in many cases. Two typical examples are illustrated in Figure 5 for different faces of the *fcc* metal rhodium. The two surfaces are the *fcc*(110) and the *fcc*(311), and the upper figure in each case shows the plan of the surface. In these plans only two layers are evident: the upper layer (layer 1) shown in green and the second layer (layer 2) in grey. To illustrate the vertical separation of the layers a different type of representation is needed; the lower part of Figure 5 shows sections through the upper layers of the surface, with layer 3 also in grey. The distances $d_{1,2}$ and $d_{2,3}$ are the separations of the layers, from which interlayer relaxation is measured. The magnitude of this interlayer relaxation is shown in Table 1, for the topmost and some lower layers. Two observations can be made from the data in Table 1.

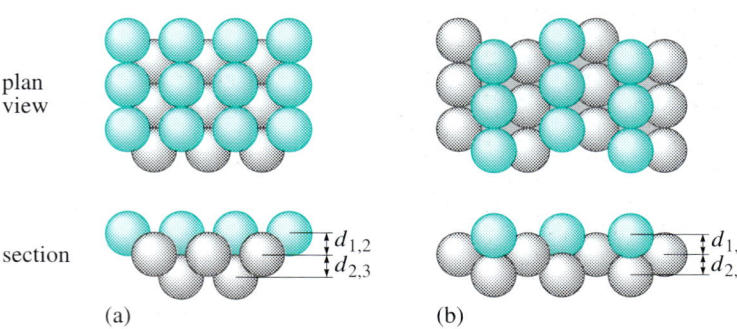

plan view

section

(a)   (b)

Figure 5 Vertical relaxation in two surfaces of the *fcc* metal rhodium: (a) the Rh(110) surface; (b) the Rh(311) surface. The upper figures are plans and the lower figures are sections through the top three layers. As in Block 6, the uppermost layer is shown in colour, and lower layers are in grey. $d_{1,2}$ is the interlayer spacing of layers 1 and 2; similarly, $d_{2,3}$ is the spacing between layers 2 and 3.

Table 1   Vertical relaxation of rhodium surfaces[a]

Relaxation	Rh(110)	Rh(311)
$\Delta d_{1,2}/d_0$	−6.9%	−14.5%
$\Delta d_{2,3}/d_0$	+1.9%	+4.5%
$\Delta d_{3,4}/d_0$	−1.9%	

[a] d_0 is the interlayer spacing in the bulk. $\Delta d_{1,2}$ is the relative difference in the interlayer spacing between layers 1 and 2 and d_0; that is, $\Delta d_{1,2} = d_{1,2} - d_0$.

First, whereas there is a contraction between layers 1 and 2, an expansion occurs between layers 2 and 3, with another contraction between layers 3 and 4. This alternation can again be explained by bonding. Layer 2 contributes to the stronger bonds between layers 1 and 2, as described above. This leaves layer 2 able to form only weaker bonds with layer 3, due to the lower electron density between these layers. Similarly, layer 3 is then left with additional electron density to form bonds with layer 4, and so on. The alternation of strength of bonding is reflected in an alternation in interlayer spacing: contracted, expanded, contracted, and so on, but the effect weakens rapidly with depth.

The second effect is the large difference in relaxation observed for the two faces. Comparison of the two surfaces (Figure 5) reveals the reason for this difference. Atoms on the (110) surface lie closer to each other than they do on the (311) surface. This is an example of the generality introduced in Block 6, that higher index planes have a more open structure than planes of lower index. Bonding between the atoms *in* the surface layer is stronger for the more closely packed surface, that is (110). Its response, by relaxation, to the absence of an overlying layer is therefore smaller. This is an example of a general observation concerning relaxation: the magnitude of the relaxation increases with the openness or **atomic roughness**, as it is termed, of the surface.

2.4   Lateral relaxation or displacement

As explained above, metals might be expected to show *small* displacements in their surface structure. In most cases, the undirected bonding in metals makes them respond to the absence of an overlayer by vertical relaxation. For about a dozen metals, atoms in the topmost surface layers have been found to be displaced *slightly* from their ideal (bulk) positions by horizontal displacement. These are examples of lateral relaxation.

An important example is found in the *bcc* metal tungsten. Because the W(100) surface can be readily cleaned, simply by heating, it is one of the most intensely studied of all surfaces; most interpretations are based on a bulk-truncated surface.

■   What would you expect the LEED pattern of the bulk-truncated W(100) surface to look like? (See Block 6, Section 8.2.2 if you are unsure about this.)

▨   For a *bcc* metal the (100) surface has a square unit mesh. So the reciprocal unit mesh, which is the repeating unit of the LEED pattern, is also square.

As Figure 6a shows, this is the observed LEED pattern at 370 K. However, a study of this surface in 1979 showed that the LEED pattern changes on cooling (Figure 6b), a change that can be interpreted as a surface restructuring.

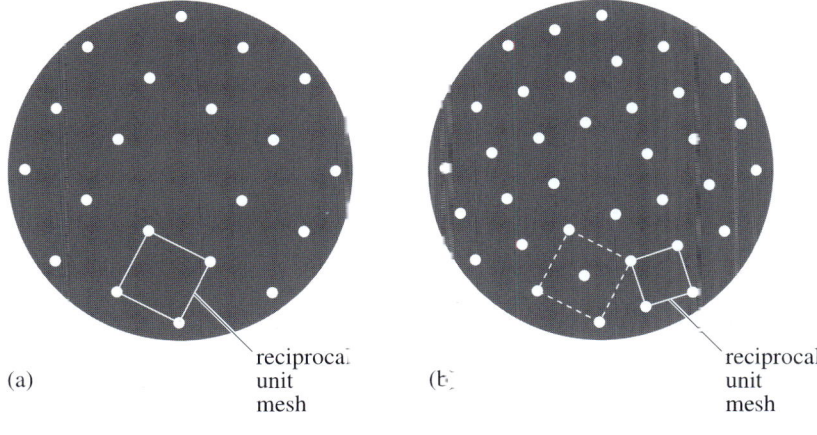

(a)   reciprocal unit mesh

(b)   reciprocal unit mesh

Figure 6   The LEED patterns of the W(100) surface: (a) at 370 K, (b) at 150 K. Note that in these patterns the full complement of spots is shown apart from that at the centre of the pattern, although in practice the presence of the sample and its holder often obscures more spots. The reciprocal mesh of (a) is reproduced as a dashed mesh in (b).

■ Notice that in Figure 6 the reciprocal mesh of the low-temperature surface is also square, although it is rotated by 45° with respect to the high-temperature form, and the spot separation is smaller. What conclusions can be drawn from this comparison? (Note that the sample orientation was the same in both parts (a) and (b).)

▪ From the relation between the surface mesh and the reciprocal mesh of a given structure, we conclude that the low-temperature surface has a real unit mesh that is both larger than and orientated differently from the high-temperature surface.

In fact, the spot separation in the LEED pattern is reduced by a factor of $\sqrt{2}$ on cooling to 150 K. So the surface unit mesh size *increases* by $\sqrt{2} \times \sqrt{2}$. An analysis of the profiles and intensities of the LEED spots has enabled a detailed structure to be deduced for the low-temperature surface. This, and the high-temperature form, which is bulk-truncated, are illustrated in Figure 7, where the unit meshes of the two structures are also shown (in black). If you look obliquely at Figure 7b from lower left to upper right of the figure, the new spatial relationship of the atoms that results from the restructuring is apparent. The atoms are arranged in zigzag chains aligned at 45° to the grid lines of the surface mesh of the high-temperature form.

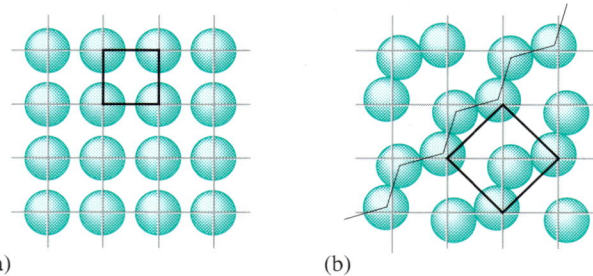

(a)   (b)

Figure 7   Plan view of the surface structures of W(100) at (a) high temperature and (b) low temperature, as deduced from LEED. In both parts of the Figure the grey lines represent the surface unit mesh of the high-temperature form. The unit mesh of each form is shown in black. The black zigzag line shows the chain structure. Displacements are exaggerated.

This dramatic change of structure is the result of a surprisingly small displacement of atoms — by a movement of ≈0.02 nm in the [110]-type direction, as shown in the top left of Figure 8, a displacement of about 4%. In Figure 8, the unit mesh of the high-temperature surface structure is shown on the left. The positions of the tungsten atoms in the low-temperature surface are shown in green, and one unit mesh is outlined in green. Displacements of the atoms are in the [110]-type direction, indicated by the arrow at top left in the Figure. The resultant chains of atoms are indicated by the black zigzag line. Notice that the surface unit mesh of the low-temperature form measures $\sqrt{2}a \times \sqrt{2}a$, and that it is rotated by 45° relative to the high-temperature surface unit mesh. So the low-temperature form is designated as W(100)($\sqrt{2} \times \sqrt{2}$)R45°, according to the notation introduced in Block 6, Section 7.2. In addition to the lateral displacement, there is a *vertical* displacement of about 6%, shortening the distance between layers 1 and 2. Atoms in the second layer are also displaced slightly in a similar zigzag manner by about 1%.

On heating, the zigzag chains break up, destroying the long-range order in the surface layer, although displacements persist locally well above room temperature. So a clean W(100) surface *never* has the ideal atomic arrangement of a bulk *bcc* (100) plane.

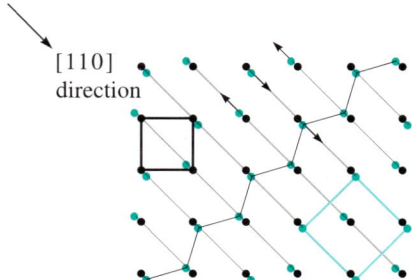

[110]
direction

Figure 8   The structure of the low-temperature form of the W(100) surface (in green) compared with the bulk-truncated high-temperature structure (in black), showing the displacement of atoms (much exaggerated) in the [110]-type direction (grey lines).

STUDY COMMENT   The relationship between the LEED patterns of the low-temperature and high-temperature structures of the W(100) surface is similar to that between the substrate and adsorbate structures in SAQ 15 of Block 6 (Section 8.3). It may help to refer back to that example if you had difficulty following the discussion above.

2.5 Clean-surface reconstruction

Reconstruction is unlike relaxation in that it involves relatively large movements of atoms, changes in surface density and the breaking of chemical bonds. For these reasons it is an activated process, in contrast to relaxation, which is often favoured by *low* temperature. As discussed above, reconstruction is expected to be more common in covalent solids, such as semiconductors, rather than metals, because of the directional nature of covalent bonding.

The classic example of reconstruction is that of silicon, which has the diamond structure (Figure 9), and reconstructs on the (111) surface, as shown by a LEED experiment. Where reconstruction occurs, the LEED pattern is complicated by the interaction of several surface layers. Although this makes the interpretation more difficult, it is sometimes possible to deduce the structure of these layers. In the (111) surface of silicon, the top three layers undergo reconstruction as illustrated in Figure 10a–c. One surface unit mesh is outlined in black in each layer in Figure 10. The unit mesh of the top layer is seven times as large along each side as that of a (111) layer of the bulk (Figure 10d), and so is designated as Si(111)(7 × 7).

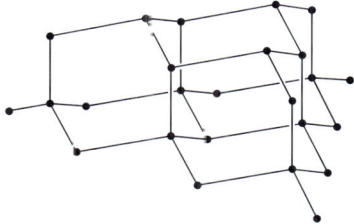

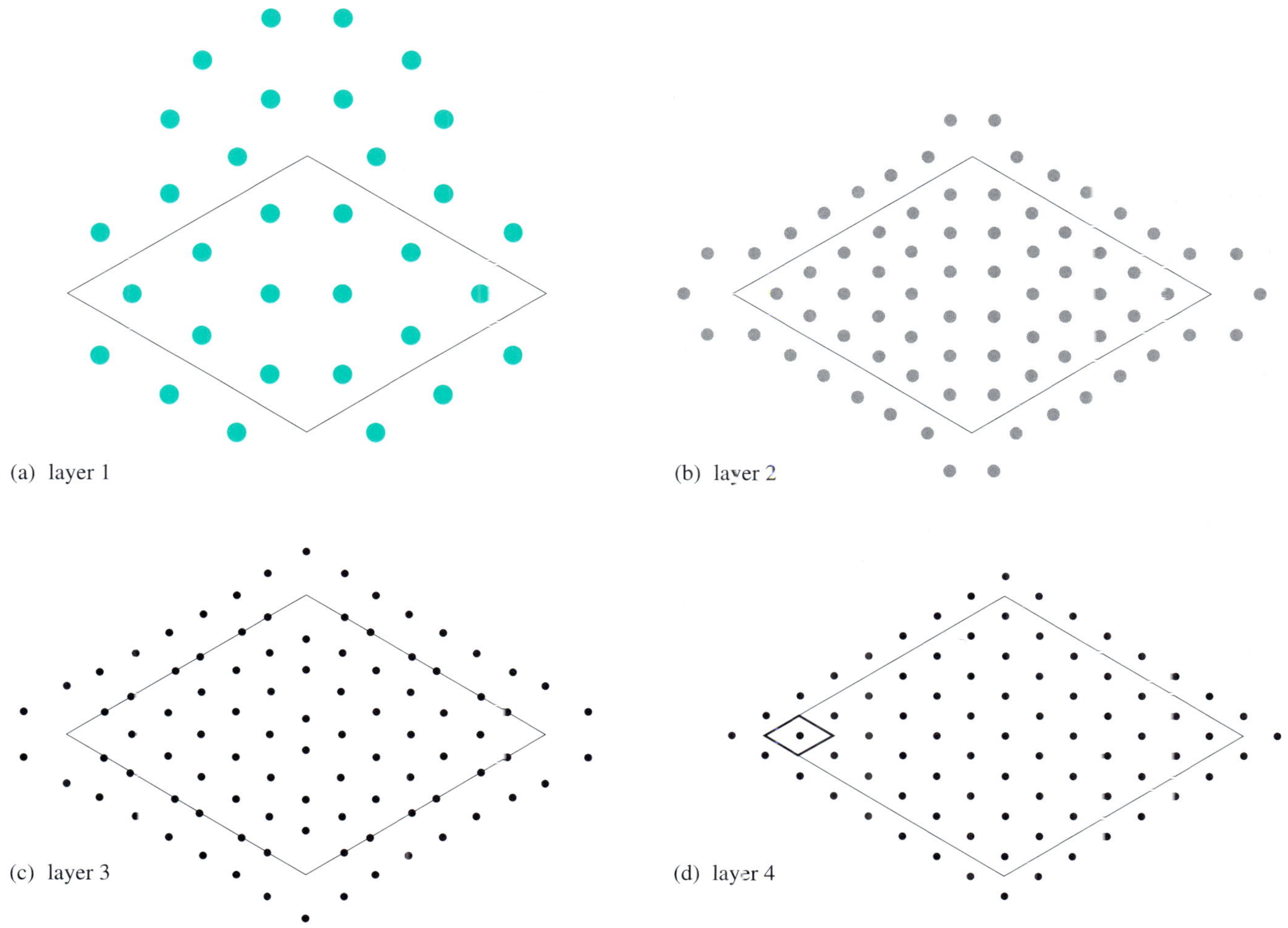

Figure 9   The diamond structure of silicon.

(a) layer 1

(b) layer 2

(c) layer 3

(d) layer 4

Figure 10   The top four layers of the reconstructed Si(111) surface: (a) the top layer; (b) the second layer; (c) the third layer; (d) the fourth layer, which has the same structure as the (111) bulk layer. The (111)(7 × 7) unit mesh is shown on each layer. The surface layer (a) is extended to show the circular pattern of atoms.

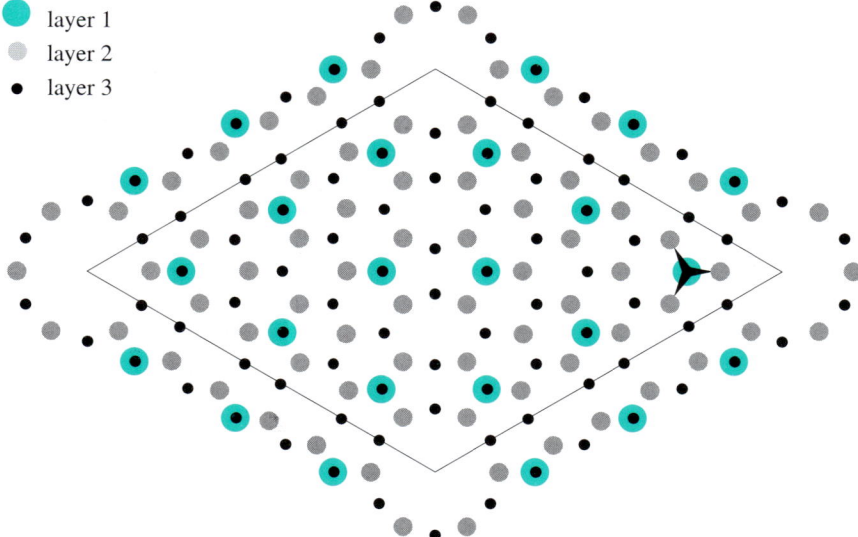

Figure 11   The top three layers of the reconstructed Si(111) surface. On the right side, one of the surface atoms is shown bonded to three atoms of the layer below, indicating that the tetrahedral environment is closely followed, although the overlying layer is absent.

When the top three layers are superimposed, the reconstructed surface structure shown in Figure 11 is generated. This structure, the deduction of a LEED experiment, has been the subject of a recent STM study. Silicon atoms were evaporated onto a (111) surface, where they formed an island (Figure 12). As the STM method images only the *top* layer, Figure 12 should be compared with the green atoms in Figure 10a. The pattern of concentric circles of atoms (6 within 12) is evident in both Figures. So the STM image strikingly confirms the results of the LEED experiment.

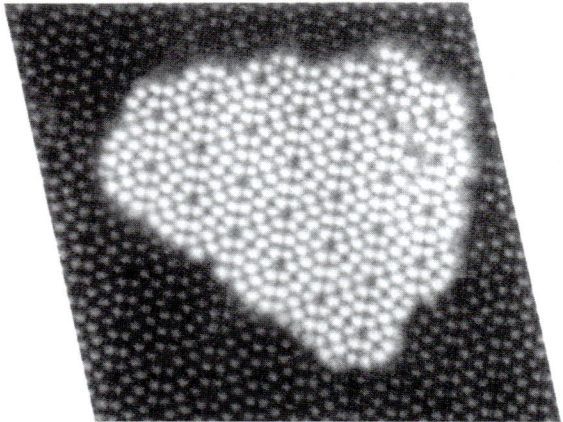

Figure 12   An STM image of an island of silicon atoms evaporated onto the Si(111) surface.

In a small number of cases, reconstruction does occur on metal surfaces, although to date the details of some of the reconstructed surfaces have not been established. Satisfactory structural analysis has been achieved, however, for the (100) and (110) surfaces of the sixth Period metals iridium, platinum and gold — all *fcc* metals. At higher temperatures, the (100) surfaces of these metals undergo a similar reconstruction, in which the atom density in the first layer is *increased* by compression into an arrangement resembling hexagonal close packing, while all other layers remain unaffected. The LEED pattern of the Ir(100) surface is shown in Figure 13.

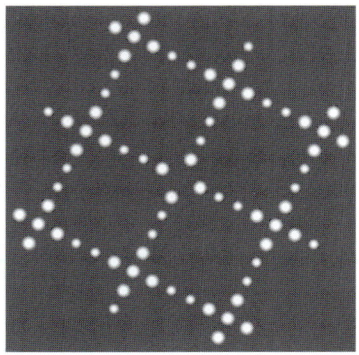

Figure 13   The LEED pattern of the Ir(100) reconstructed surface.

■ How does the pattern in Figure 13 appear modified from your expectation of that for a *fcc*(100) surface (depicted in Figure 14)?

▨ Instead of a (1 × 1) square pattern, this square pattern repeats every five spots. Note that the spots have various intensities.

An analysis of this pattern, by comparison with models, reveals that the surface layer is as shown in Figure 15a, a perspective view in which the two rows at the left and right of the top (green) layer indicate the repeat pattern in the *combined* layer 1–layer 2 assembly. Section and plan views are also shown in Figure 15. It is evident that reconstruction has occurred.

■ Use Figure 15 to express the repeat distance between the (heavily outlined green) rows in the reconstructed surface in terms of that in the 'bulk' (100) plane (layer 2).

▨ Comparison of the outlined green rows in layer 1 with the rows in layer 2 shows that the repeat distance in layer 1 is five times that in layer 2.

It is also apparent from the perspective view in Figure 15a that there is no mismatch of repetition *along* the rows in layers 1 and 2. Hence, the reconstructed surface may be labelled Ir(100)(1 × 5), according to the ($m \times n$) notation in Block 6 (Section 7.2). As the bulk-truncated structure has unit mesh sides of equal length ($a = b$ in Figure 14), an equivalent notation for the reconstructed surface is Ir(100)(5 × 1). It is evident from the section shown in Figure 15b that the packing of six surface-layer atoms on top of five bulk atoms is achieved by buckling of the surface layer.

In these same metals a more severe form of reconstruction is observed in the (110) surface. Again, the surface structure has been deduced mainly from LEED studies and by modelling both the pattern and the intensities of the spots. The structures that emerge from these studies contrast with those of the (100) surfaces in two ways. Firstly, the top three layers are reconstructed and vertical relaxation also occurs. Secondly, the main feature of the reconstruction is a *decrease* in the atom density of layer 1 caused by the apparent loss of alternate rows, giving the structure shown in

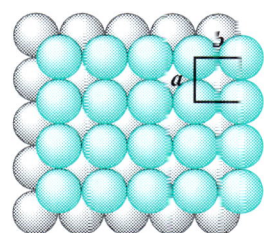

Figure 14   The unreconstructed *fcc*(100) surface The top layer is shown in green and the second layer in grey. The surface unit mesh is outlined in black. The unit mesh vectors have magnitudes $a = b$.

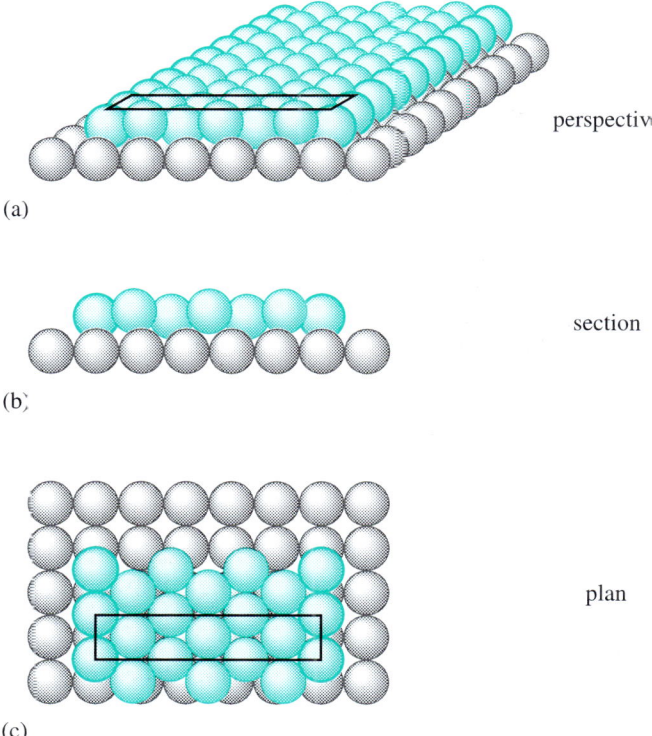

perspective

(a)

section

(b)

plan

(c)

Figure 15   The structure of layers 1 and 2 in the reconstructed (100) surface of iridium: (a) perspective view; (b) section view; (c) plan view, showing a surface unit mesh. (Note the hexagonal close-packed appearance of the top layer as shown in part (c).)

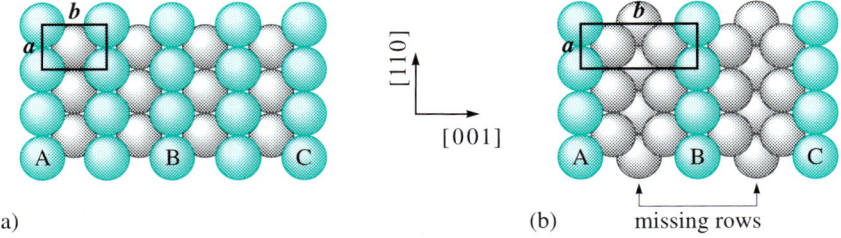

(a)   (b)   missing rows

Figure 16 Plan views of: (a) the unreconstructed bulk-truncated structure of the Pt(110) surface; (b) the missing row (MR) reconstruction of the Pt(110) surface. The surface unit mesh is indicated in each case.

Figure 16b. This is called a **missing-row (MR) structure**. Comparison of the MR structure (Figure 16b) with the bulk-truncated surface (Figure 16a) shows that the unit mesh is the same size in the *a* direction but double in the *b* direction. Hence the MR surface is denoted as (1×2). Notice that in this case the unit mesh sides are unequal $(a \neq b)$, and so this surface is *not* equivalent to the (2×1) surface.

◼ By comparing Figure 16a and Figure 16b, suggest how the atoms in layers 2 and 3 respond to the missing rows of layer 1.

▨ The atoms in layer 2 are displaced laterally in the [001]-type directions to form alternating pairs of rows. In layer 3 there does not appear to be any lateral displacement, but note that in Figure 16b alternate rows of layer 3 are obscured by the atoms in layer 1.

However, vertical relaxation occurs in layers 1 and 3, according to the sketch in Figure 17, which exaggerates these movements in comparison to the 'ideal' bulk structure. Notice that layer 1 is displaced downwards while layer 3 is buckled, the main relaxation being an upwards displacement of alternate rows into the space of the missing rows in layer 1, as indicated by the arrows.

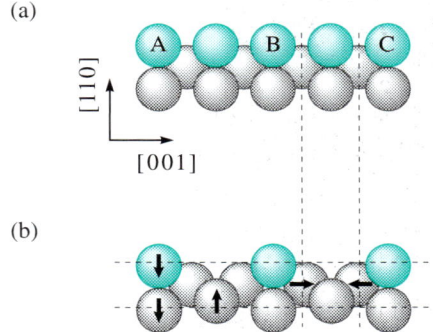

Figure 17 An end-on view (of the edge ABC in Figure 16) through the top three layers of the (a) bulk-truncated and (b) MR reconstructed (110) surface of platinum. Vertical relaxation is exaggerated. Note the missing alternate rows in the top layer of the MR surface.

2.6 Summary of Section 2

1 Surface restructuring describes the situation in which the atoms on a crystal surface occupy different positions from those expected from the ideal bulk structure.

2 Inaccurate cleaving of crystals and point defects prevent the attainment of perfect crystal surfaces.

3 Interlayer relaxation involves the vertical movement of whole layers, usually with alternating increased and decreased interlayer spacing.

4 Lateral relaxation or displacement involves the small horizontal movement of atoms within layers. Small movements may create chain-like structures.

5   Reconstruction involves relatively large movements of atoms from their ideal positions. It may involve bond breaking and making, and is more common in covalent solids. In extreme cases, reconstruction results in surfaces with missing rows.

SAQ 1   The (110) surface of the *fcc* metal rhodium shows interlayer relaxation (Figure 5 and Table 1). Hydrogen adsorbs dissociatively on this surface, and with increasing coverage the interlayer relaxation decreases, tending towards the spacings of the bulk. Suggest how the effect of hydrogen adsorption on the bonding between surface layers provides a plausible explanation for this observation.

SAQ 2   Nickel and platinum have *fcc* structures. How would you expect the different atom densities of the two surfaces, Ni(111) and Pt(210), to affect interlayer (vertical) relaxation? These surfaces are shown in Figure 18.

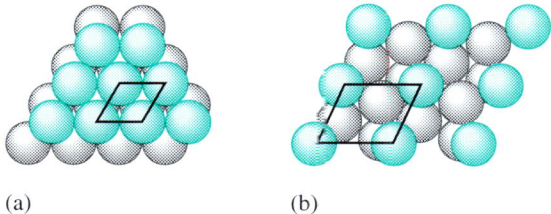

(a)   (b)

Figure 18   (a) The (111) surface of nickel; (b) the (210) surface of platinum.

SAQ 3   If the reconstructed (100) surface of iridium is exposed to oxygen and flash heated to 750 K, the LEED pattern changes from that in Figure 19a to that in Figure 19b. What do you think has happened to the surface in this process?

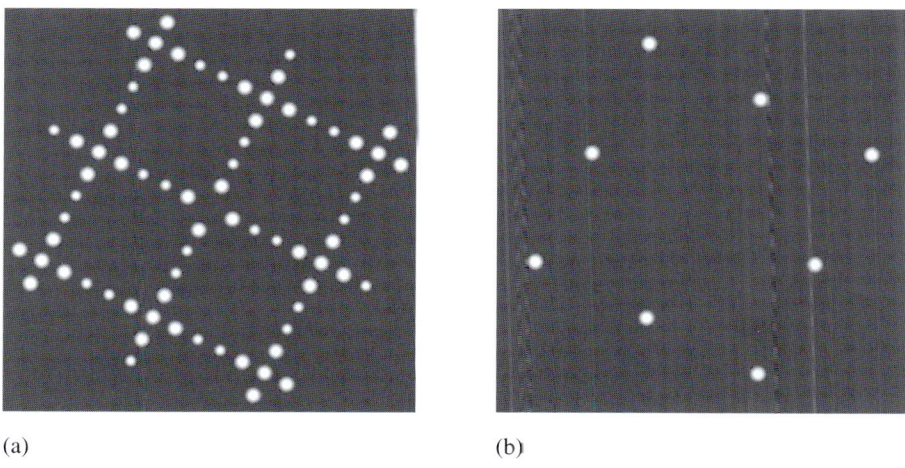

(a)   (b)

Figure 19   The LEED pattern (a) of the reconstructed Ir(100) surface, and (b) of this surface after treatment with oxygen at 750 K.

3 ADSORBATE-INDUCED PROCESSES

As surface structure responds to the absence of overlying layers, so too does it often respond to the adsorption of atoms and molecules. This is less surprising than the phenomenon of clean-surface reconstruction; the arrangement of surface atoms should adapt to the adsorbed species just as the adsorbate adapts to the surface (as in dissociative adsorption). In general, weak adsorption results only in small displacements, whereas chemisorption produces mass transport and reconstruction. Being an activated process, reconstruction is seen most dramatically at high temperatures, an observation that has important consequences for the interpretation of catalytic studies.

Information about the effects of adsorbates on surface structure comes mainly from LEED experiments. As LEED patterns are determined by the top few layers of a surface, it is possible by this method to deduce the structure of the adsorbate layer and the underlying substrate.

3.1 Adsorbate-induced relaxation

You will recall the lateral relaxation observed by LEED in the low-temperature W(100) surface, in which the surface atoms move in the [110]-type direction to create a top layer of zigzag chains. This structure was illustrated in Figures 7 and 8, and is repeated here as Figure 20a.

When this surface is exposed briefly to hydrogen at room temperature, so that the fractional surface coverage is not more than about 0.25, pairs of surface tungsten atoms rotate slightly (Figure 20b) to form dimers, thereby breaking the zigzag chains. The dimers so formed provide the sites for adsorption of hydrogen. In Figure 20, surface tungsten atoms are shown in green, indicating how the rotation of the pairs occurs about an axis perpendicular to the page.

■ Compare the positions of the tungsten atoms in the top layer in Figure 20b with the positions of the tungsten atoms in the top layer of the unreconstructed surface in Figure 20c. How do the displacements that produce the dimer structure in Figure 20b compare with those that produce the zigzag chain structure in the clean surface (Figure 20a)?

■ The dimer structure results from displacements in the [100]-type direction, whereas the zigzag chain structure results from [110]-type displacements.

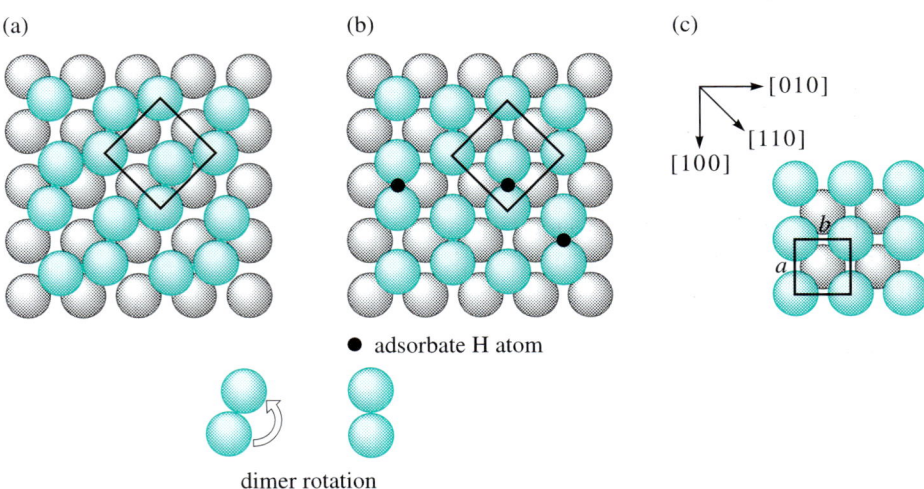

● adsorbate H atom

dimer rotation

Figure 20 (a) The modelled W(100) surface: the clean low-temperature W(100)($\sqrt{2} \times \sqrt{2}$)R45° surface, showing relaxation-produced zigzag chains; (b) the hydrogen-induced ($\sqrt{2} \times \sqrt{2}$)R45° structure of W(100); (c) the unreconstructed W(100) structure. Notice that in Figure 20b the unit mesh is defined by the topmost tungsten layer and not by the adsorbate distribution.

Clearly, it is the twofold coordination of hydrogen atoms that acts as the driving force for the relaxation that converts the chain structure to the dimer structure. As one hydrogen atom adsorbs at one dimer site, Figure 20b seems to suggest that a fractional coverage of 0.5 by hydrogen would provide greatest stability for the dimer surface. However, the dimer structure of Figure 20b is established at a fractional coverage of 0.25. As the hydrogen adsorption at this coverage is not ordered, the surface unit mesh is still defined by the top layer of tungsten atoms, as indicated in Figure 20b. In fact, at fractional coverages greater than about 0.3, this dimer structure gives way to a different phase, as it also does with increased temperature. Predictions based on such simple structural considerations are rarely observed.

STUDY COMMENT   We have made use of the $(m \times n)$ notation (see Block 6, Section 7.2) in denoting structures produced by clean-surface restructuring (see Section 2.4) and by adsorbate-induced processes (as discussed above). The following SAQ provides further examples. It also enables you to check that you can determine the fractional surface coverage for a given structure. Refer to Section 7.3 of Block 6 if necessary.

SAQ 4 (revision)   At low temperature $(T < 200 \text{ K})$, hydrogen adsorbs on the (110) surface of the *fcc* metal nickel in a way that depends on the coverage. At a fractional surface coverage of $\theta \leqslant 1.0$, the hydrogen atoms occupy the sites labelled X in Figure 21a. At higher coverage the rows of nickel atoms relax, closing and opening adjacent gaps (Figure 21c and d), in a structure that is called **row paired**, **RP** (see also Section 3.2). In this structure, the hydrogen atoms occupy two different types of site, for example those labelled Y and Z, where the Z site is similar to the X site at low coverage.

For each of the structures in Figures 21a and 21c, identify the surface unit mesh. Label the mesh using the $(m \times n)$ notation, and determine the number of adsorbed species contained within it. Hence, calculate the fractional surface coverage, θ, for each structure.

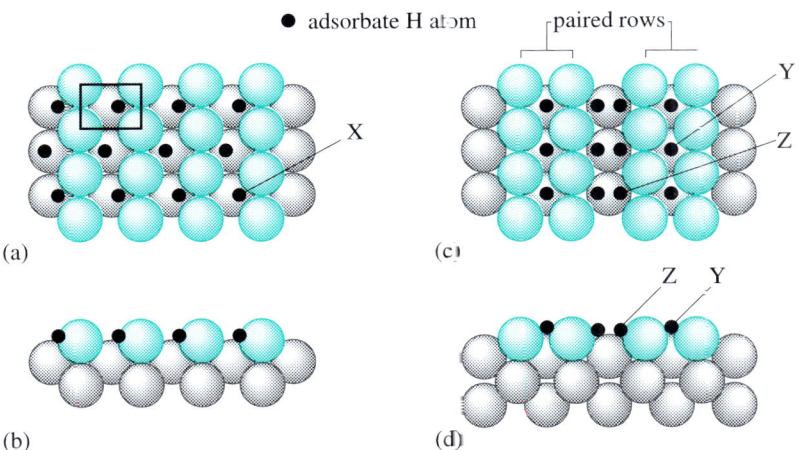

(a)   (b)   (c)   (d)

Figure 21   LEED-deduced structure produced by hydrogen adsorption on Ni(110), at low coverage (a and b) and high coverage (c and d), showing plan views (a and c) and section views (b and d), respectively. The unit mesh of the 'ideal' Ni(110) surface is indicated in (a).

The examples in SAQ 4 raise a point about notation that was not covered explicitly in Block 6. In each of the structures in Figure 21, the surface mesh contains more than one atom of the adsorbed species, namely 2H in Figure 21a and 3H in Figure 21c.

The surface meshes are therefore denoted as Ni(110)(2 × 1)–2H and Ni(110)(1 × 2)–3H, respectively.

3.2 Adsorbate-induced reconstruction

Following the example of hydrogen adsorbed on the Ni(110) surface in SAQ 4, a recent LEED study shows that reconstruction occurs at temperatures greater than 200 K. The LEED data have been interpreted in terms of a surface in which a (1×2) missing-row structure prevails, with hydrogen atoms occupying three-coordinate sites (labelled X in Figure 21). The plan and section views of this structure, reminiscent of the clean-surface reconstruction of Pt(110) (Figure 16 in Section 2.5), are shown in Figure 22. Essentially, the surface consists of a MR structure of Ni—H—Ni chains. The unit mesh (indicated in Figure 22a) contains two hydrogen atoms, and so the structure is denoted as Ni(110)(1×2)–2H.

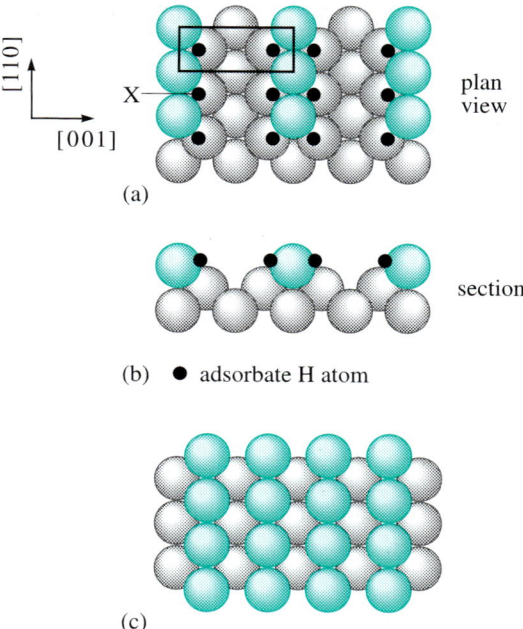

(a)

(b) ● adsorbate H atom

(c)

Figure 22   The LEED-deduced structure of the Ni(110)(1×2)–2H surface above 200 K; (a) plan, and (b) side view. For comparison, the clean unreconstructed Ni(110) surface is shown in (c).

Again, interpretation of the LEED results is based on modelling of the pattern and intensities of the spots. In this case, however, stunning supporting evidence has been obtained from a STM study of this surface. The STM image — with atomic resolution — of the *clean* Ni(110) surface (Figure 22c) is shown in Figure 23a. Brief exposure to hydrogen at room temperature shows (Figure 23b) the formation of isolated chains composed of atoms that lie along the [110]-type directions of closest packing. We are now in a position to compare the results of the LEED and STM experiments on the Ni(110) surface that is reconstructed by hydrogen adsorption. In the LEED-deduced structure in Figure 22a, the separation of the rows of nickel atoms has been measured as 700 pm.

■   What is the separation of the rows in the clean Ni(110) surface shown in Figure 23a? The STM frame represents a piece of surface 2 100 pm × 2 100 pm.

▨   In Figure 23a there are about 7.5 spacings on the diagonal of the image. The diagonal measures $\{\sqrt{(2\,100^2\,\text{pm}^2 + 2\,100^2\,\text{pm}^2)}\} = 2\,969.8$ pm, so the separation of the rows is $(2\,969.8/7.5)$ pm = 396 pm.

The separation of the rows in the top layer of the MR structure (Figure 22) is expected to be twice this value. As the LEED experiment shows, the two techniques give results in good agreement. In fact, a local MR structure is shown imaged by STM in Figure 23c. Here the separation of adjacent rows in the top layer is about 750 pm, supporting the result of the LEED experiment. While the STM image confirms the LEED model of chains of nickel atoms, it offers no direct support for the hydrogen atoms at X-sites because these are not visible by STM.

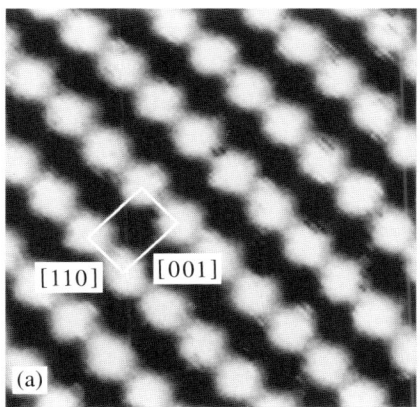 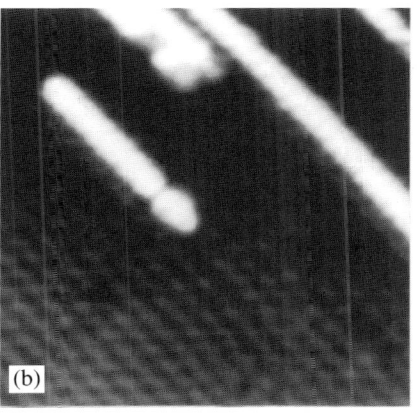 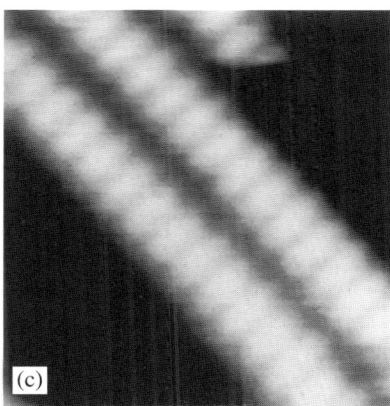

Figure 23   The result of a STM study of the Ni(110) surface on exposure to hydrogen:
(a) the bare surface, 2 100 pm × 2 100 pm; (b) the surface exposed to hydrogen, 7 100 pm
× 7 500 pm; (c) an image showing adjacent Ni—H—Ni chains, 3 100 pm × 3 200 pm.

In addition to providing images of static surfaces, STM provides the possibility of
studying the structures of surfaces as they change *during* adsorption. Such dynamic
studies by STM reveal details of the growth of a surface that are not evident from
a single observation. For example, in the hydrogen adsorption on Ni(110) described
above, it is observed that on terraces that are far removed from any steps or other
imperfections the (1 × 2) reconstruction occurs by removal of rows of nickel atoms
from the flat terrace, creating missing rows. In turn, these nickel atoms are deposited
on other parts of the terrace to form **added rows. AR**, lying along [110]-type directions. Initially, the distribution of these missing and added rows is quite random,
as Figures 23b and c indicate, but further exposure to hydrogen creates an ordered
surface with the rows spaced uniformly by about 750 pm, as described above.

An important conclusion from this type of study is that at high coverages, the
MR and AR descriptions are identical, and so the terms are relevant only to the
mechanism by which the surface forms. This means that a distinction between them
cannot be made by static studies alone, but requires dynamic STM observations of
the way in which the overall structure evolves with increasing coverage.

The missing row structure with metal–adsorbate–metal chains, described above for
hydrogen on Ni(110), turns out to be quite common in surface studies. An interesting
example that illustrates the changes imposed by increasing coverage is that of oxygen
on the Ni(110) surface. STM images at different oxygen coverages are shown in
Figure 24; as with hydrogen, the oxygen atoms are not visible. In each case the
surface is characterized by chains of nickel atoms running in the [001]-type
directions; note the difference from the case of hydrogen adsorption, where the
Ni—H—Ni chains lie along the [110]-type directions (Figure 23), as do the close-
packed rows on the clean Ni(110) surface (evident in Figure 22c).

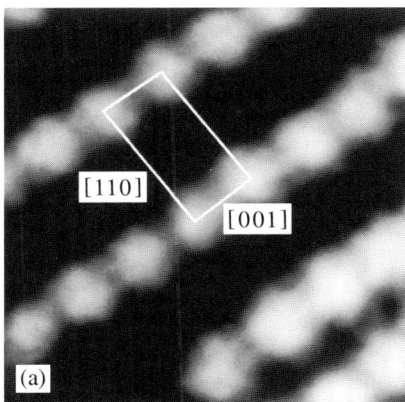 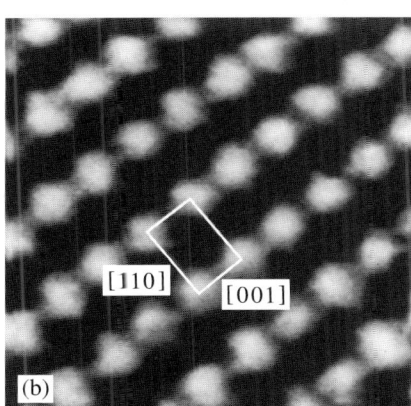 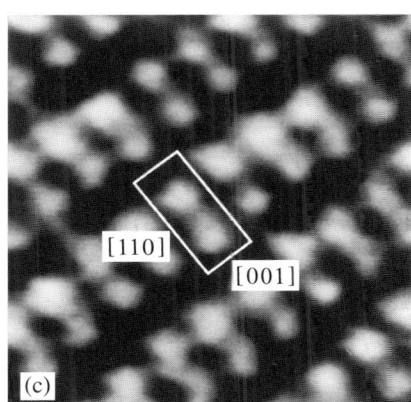

Figure 24   STM images of an area of the Ni(110) surface at different coverages of oxygen:
(a) $\theta = \frac{1}{3}$; (b) $\theta = \frac{1}{2}$; (c) $\theta = \frac{2}{3}$. In each case a unit mesh is shown.

The interpretation of these STM images is depicted in Figure 25, which also shows the clean Ni(110) surface. In all three cases the structures are of nickel–oxygen chains, directed as shown in Figure 25 along the [001]-type direction. In Figures 24c and 25d you will notice that adjacent rows of nickel atoms have moved closer to each other, to give what can be described as a row-paired RP structure (see SAQ 4 in Section 3.1). Although these structures and the STM images appear quite static, at room temperature the chains are mobile, showing a tendency to aggregate into local regions (islands) of one or other type, depending on the coverage.

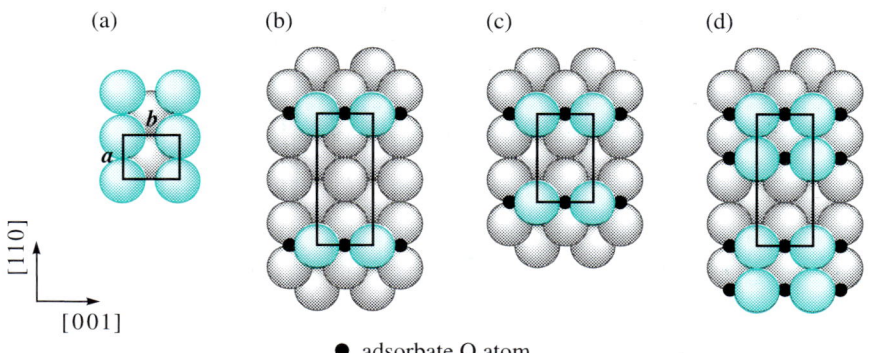

● adsorbate O atom

Figure 25   The local structure of Ni—O—Ni chains as deduced from the STM images in Figure 24: (a) the clean unreconstructed surface, (b) the (3×1)—O structure for $\theta = \frac{1}{3}$, (c) the (2×1)—O structure for $\theta = \frac{1}{2}$, (d) the (3×1)—2O structure for $\theta = \frac{2}{3}$. The surface unit mesh is indicated in each case.

The above examples are but a few of the many adsorbate–metal chains of MR and AR types that have been observed for a variety of metals and adsorbates, particularly hydrogen, oxygen and sulphur.

3.3 Summary of Section 3

1   Surfaces respond to the presence of adsorbates by restructuring.

2   Generally, weak adsorption results in small displacements. For example, hydrogen adsorption on W(100) results in relaxation, converting the zigzag chain structure to a dimer structure.

3   Strong adsorption (chemisorption) results in large displacements (reconstruction). For example, hydrogen adsorption on Ni(110) generates a missing-row (MR) structure.

STUDY COMMENT   Check that you agree with the labels and fractional surface coverages given in the caption to Figure 25. Then try the following SAQ.

SAQ 5   Oxygen interacts with Cu(110) in a similar way to its interaction with Ni(110). The following experimental observations have been made. In regions close to a step edge the copper required to form Cu—O chains is supplied by the removal of copper atoms from *along* the step edge. On terrace regions far from steps, it is supplied by the creation of troughs one atom layer deep. Decide whether these two cases are missing-row or added-row processes.

4   RESTRUCTURING AND CATALYSIS

From the previous Sections we conclude that single-crystal surfaces respond to changes in temperature and to the presence of adsorbates with some degree of dynamic structural change. We may also predict that the same will be true of catalysts, which, even in a finely divided form, often have typical crystal faces.

In this Section, we consider examples relating to two industrially important catalytic reactions — alkane hydrogenolysis and alkene hydrogenation. We end by outlining a generalization, which, although tentative, provides an explanation for a long-standing catalytic puzzle.

4.1   Structure-sensitive and structure-insensitive reactions

The rates of many catalysed processes are found to depend on the structure of the surface. For example, a given reaction often proceeds much more rapidly on one type of crystal surface than on another surface of the same catalyst. However, some reactions are found to be **structure-insensitive**; that is, they appear to occur at the same rate on *different* single crystal planes of a particular metal. An indication that, in certain cases at least, this insensitivity may be a consequence of the conditions under which the catalyst is prepared and/or operated is provided by a study of the rhodium-catalysed hydrogenolysis of pentane (equation 1). This reaction proceeds by two paths, according to which C—C bond is broken.

$$C_5H_{12} + H_2 \begin{cases} CH_4 + C_4H_{10} \\ \\ C_2H_6 + C_3H_8 \end{cases} \tag{1}$$

On clean rhodium this reaction is **structure sensitive**. For example, the nature of the products and the activation energies differ over Rh(100) and Rh(111) surfaces, indicating different mechanisms (Table 2).

Table 2   Data for the rhodium-catalysed hydrogenolysis of pentane

Surface	C—C bonds broken	Products	E_a/kJ mol^{-1}
clean Rh(100)	terminal	methane, butane	90
clean Rh(111)	inner	ethane, propane	200

When, however, the two surfaces are first oxidized by oxygen and then reduced in hydrogen at 400 K — a temperature sufficient for rhodium oxide decomposition but not for sintering of rhodium metal — the activation energy for subsequent hydrogenolysis is ≈ 90 kJ mol^{-1} on *both* surfaces. This does not mean, however, that either is now like clean Rh(100), for both show a selectivity shift in favour of multiple C—C bond breaking, to give methane as the sole product. Thus, the oxidation–reduction process must produce a completely new type of active site *which is the same in both the (100) and (111) cases*. In other words, the structure sensitivity of clean rhodium has been eliminated by restructuring.

Oxidation leads to an increase in volume because oxides have lower densities than metals. Therefore, when the oxygen is removed again, the solid may be left in a reconstructed state if the reduction temperature is not sufficiently high to ensure efficient sintering of the metal particles.

Catalyst pre-treatment of this kind is often used industrially to adjust the reactivity of supported metals. Exactly the same effects have been observed with a silica-supported rhodium catalyst. After pre-oxidation and annealing in hydrogen at *high* temperature (773 K), where sintering *would* occur, 5–10 nm rhodium particles with predominantly (111) facets were formed. Their behaviour in pentane hydrogenolysis was similar to that of the clean Rh(111) single crystal discussed above. On the other hand, when the oxidized catalyst was reduced at low temperature (400 K), the particles were converted to polycrystalline aggregates with roughened surfaces and no perfect facets. As with the two single crystals treated in this way, hydrogenolysis then produced mainly methane with an activation energy of $\approx 85\,\text{kJ mol}^{-1}$.

4.2 Dynamic recycling of active sites

The discussion up to this point might leave the impression that restructuring is an isolated event in the history of a catalyst. However, it could well be a continuous and very rapid process. An indication of this possibility is given by a simple adsorption system. Figure 26 shows the STM image of a Au(111) surface exposed to sodium vapour ($\theta_{\text{Na}} = 0.15$). In the centre of the picture two terraces are separated by a step of height one gold atom. The first thing to notice is the absence of any direct evidence of sodium atoms. However, their presence is apparent from a comparison of this image with that of the same surface region when sodium-free. In that case, the step appears quite sharp, whereas here it is fuzzy. These observations suggest that sodium adsorption weakens the bonding of gold atoms at the step edge to the extent that they become highly mobile. The step then undergoes a rapid and continuous restructuring on a time-scale shorter than that taken by the STM tip to scan the surface, and so the image in this region appears fuzzy.

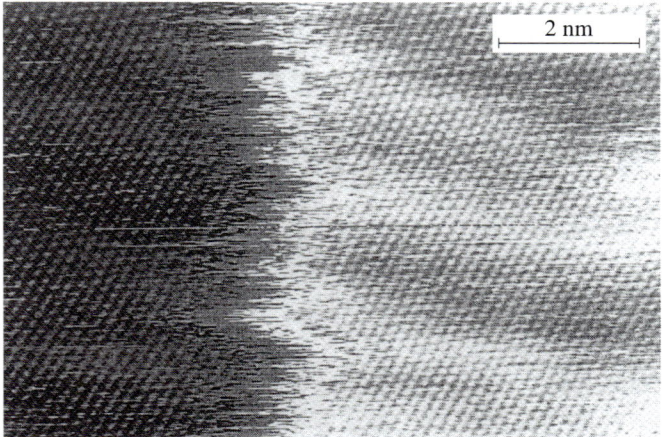

Figure 26   The STM image of a Au(111) surface with $\theta_{\text{Na}} = 0.15$.

For an example of a similar phenomenon in catalytic chemistry we return to the familiar Ni(110) surface, on which a detailed study has been made of the conversion of ethene to ethane. For reasons that do not concern us here, the isotope deuterium was used in both the hydrogen and the ethene. The overall reaction is

$$C_2D_4 + D_2 = C_2D_6 \tag{2}$$

The Ni(110) surface is unusual in its interaction with hydrogen due to the complexity of the process involved, some aspects of which were considered in Section 3.2. At 80 K, and with coverages up to one monolayer ($\theta \leqslant 1$), the LEED pattern shows that no reconstruction occurs. The surface retains the (1×1) symmetry of the underlying structure, and hydrogen occupies X sites (see Figure 21a). At fractional coverages in excess of 1, however, row-paired reconstruction occurs and hydrogen occupies both Y and Z sites (Figure 21c). When any one of these surfaces is exposed to ethene-d$_4$ (C_2D_4) and the temperature is raised, the thermal desorption spectrum that is obtained can be used to follow the progress of the catalytic reaction (equation 2 above).

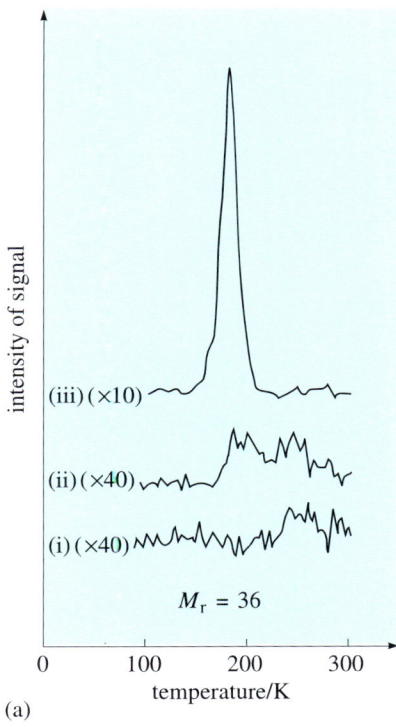

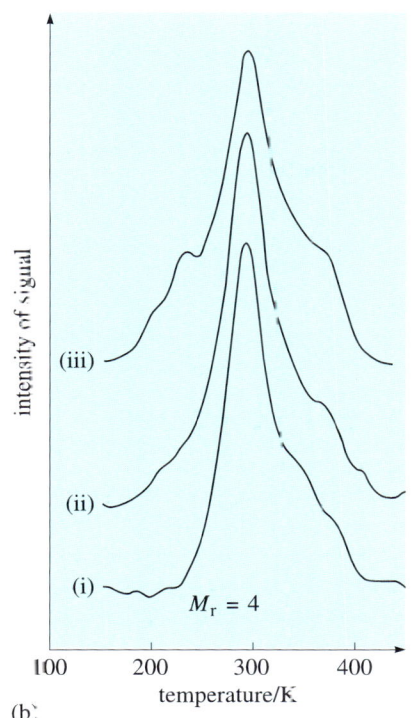

Figure 27 The thermal desorption spectra of (a) ethane-d_6 and (b) deuterium from a Ni(110) single crystal surface that has been pre-exposed first to D_2 and then to C_2D_4 at 80 K. Three spectra are shown at different fractional pre-coverages of deuterium: (i) $\theta_D = 0.70$, (ii) $\theta_D = 0.99$ and (iii) $\theta_D = 1.20$. The vertical axis indicates the intensities of the signals generated by desorption as the temperature is raised. The two peaks are at $M_r = 36$ due to C_2D_6 and at $M_r = 4$ due to D_2. Note the amplification factor needed to display the C_2D_6 peaks.

Figure 27 shows the thermal desorption spectra resulting from the C_2D_4 exposure of Ni(110) surfaces that had been previously exposed to various coverages of D_2. Mass spectroscopy used in conjunction with TDS identifies the desorbed species as either D_2 (with relative molecular mass, $M_r = 4$) or C_2D_6 (at $M_r = 36$). At low deuterium pre-coverage ($\theta_D < 1$) the surface is unreconstructed, and a very broad weak C_2D_6 peak is seen to desorb at about 250 K, together with a D_2 peak at 300 K. When, however, the fractional deuterium pre-coverage is in excess of 1 ($\theta_D > 1$), so that the surface adopts the row-paired structure, a *very much larger* C_2D_6 peak occurs at a *lower* temperature, around 180 K. The overall reaction therefore occurs much more readily on the RP surface, and this is thought to be due to the presence of deuterium at the Y sites. Why should this deuterium be released to react with adsorbed C_2D_4 more easily than deuterium at the X sites on the unreconstructed surface?

A clue to the answer is to be found in a detailed kinetic study and its interpretation. A two-step mechanism is proposed, involving sequential combination of a molecule of C_2D_4 with D atoms at Y sites, which in the process are converted to vacant X sites:

$$C_2D_4(a) + D(a,Y) \longrightarrow C_2D_5(a) + X \tag{3}$$

$$C_2D_5(a) + D(a,Y) \longrightarrow C_2D_6(a) + X \tag{4}$$

The kinetic study indicates that these reactions have higher rates than would the equivalent steps involving D at X sites, such as

$$C_2D_4(a) + D(a,X) \longrightarrow C_2D_5(a) + X \tag{5}$$

Interestingly, the activation *energies* for steps 3–5 turn out to be very similar; the reason for the rate difference can be traced to an *entropy* effect. From other experiments it is known that deuterium atoms on the surface in the vicinity of Y sites are *less mobile* and so have *lower entropy*, than deuterium at X sites. This means that when the activated complex for step 3 (or step 4) is formed from D(a,Y) as a reactant,

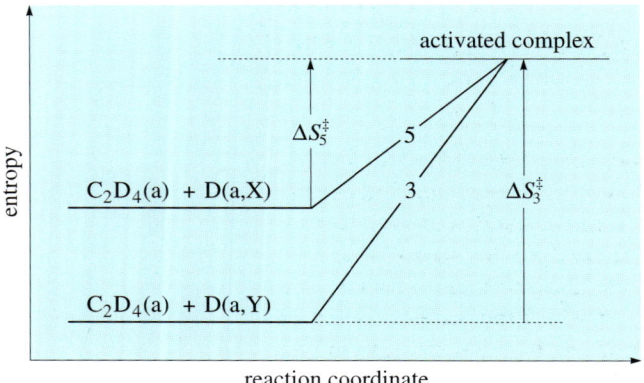

Figure 28   Comparison of the entropy of activation ($\Delta S^{\ddagger}$) for step 3 and step 5. Note the effect of $\Delta S^{\ddagger}$ on $\Delta G^{\ddagger}$, and hence its effect on the rate (see text).

there will be a *larger* entropy increase, $\Delta S^{\ddagger}$, than when starting from D(a,X) in step 5 (Figure 28). As with any process, the higher entropy increase will make activated-complex formation more favourable (since $\Delta G^{\ddagger} = \Delta H^{\ddagger} - T\Delta S^{\ddagger}$), and hence will increase the rate, for steps 3 or 4 than for steps such as 5. So, here we have a catalytic process in which the predominant driving force is an *entropy* effect, resulting from a *local* change in substrate structure (actually a conversion of a Y site to an X site).

Although Y sites are destroyed by reactions 3 and 4, a high level of reactivity is maintained by a further process, ethene *de*hydrogenation, which is found in parallel with hydrogenation. This takes place *only at X sites* and converts them back to the Y-type, ready to take part in steps 3 and 4:

$$C_2D_4(a) + X \longrightarrow C_2D_3(a) + D(a,Y) \tag{6}$$

In this example, therefore, the Ni(110) surface structure at a local level is in a highly *dynamic* state, with individual sites being recycled continuously between less active X and more active Y forms.

4.3   Catalysis on rough surfaces

Most catalytic reactions involve the breaking of bonds in one or more adsorbed reactants to give intermediates from which the products are formed. Since the early 1970s it has been known that atomically rough surfaces, which contain kinks and steps, are highly active in causing bond dissociation. It follows that rough surfaces should be especially active in catalysis. However, the fragments produced by dissociation are particularly strongly adsorbed at the very sites at which they are generated, namely at the kinks and steps. This is clearly not a favourable situation for catalysis, which requires the easy release of a fragment for binding to some other adsorbed species in order to generate a molecule of product.

One way of reconciling these apparently contradictory observations is based on the further observation that the rougher the surface the easier is restructuring, and even easier the stronger any adsorption on the rough surface. The suggestion then is that reactant dissociation at, say a kink, triggers *a localized restructuring to give a new transient site of a type at which fragment release is easy*. It is here that the catalytic step actually occurs. Afterwards, the product molecule desorbs, the transient site is destroyed, and the original kink site is regenerated.

If this model turns out to be realistic, dynamic site recycling by means of continual reconstruction and deconstruction will be seen to have significance well beyond the $H_2/C_2D_4/Ni(110)$ system discussed above. It could be one of the major factors which enable metal surfaces in general to act as heterogeneous catalysts.

5 POSTSCRIPT

As recently as 1980, in spite of the existence of many of the surface techniques described in Block 6 and illustrated here, the studies of surface science and catalysis were widely recognized as two different disciplines. In this short study we have presented a few of the many examples that demonstrate the convergence of the two subjects. The examples chosen show that a catalyst surface, or even a single crystal, can no longer be regarded as a static checkerboard on which molecules come and go. Just as the adsorbates respond, by reaction, to their location on a surface, so too does the surface respond locally to the presence of the adsorbates. This new science of dynamic restructuring is now at the forefront of surface science, and it promises to reveal valuable information about the catalytic process.

OBJECTIVES FOR TOPIC STUDY 2, PART 2

Now that you have completed Topic Study 2, Part 2, you should be able to do the following things:

1   Define, recognize valid definitions of, and use in a correct context, the terms, concepts and principles printed in bold type in the text and collected in the following Table.

List of scientific terms, concepts and principles used in Topic Study 2, Part 2

Term	Page no.
adsorbate-induced relaxation	16
added rows, AR	19
atomic roughness	9
bulk-truncated structure	8
clean-surface reconstruction	11
dynamic recycling (of active sites)	22
interlayer (vertical) relaxation	6
lateral relaxation (displacement)	6
missing-row (MR) structure	14
point defect	7
reconstruction	6
relaxation	6
row-paired (RP) structure	17
structure-insensitive reaction	21
structure-sensitive reaction	21
surface restructuring	5

2   Use simple bonding descriptions to account for interlayer relaxation. (SAQs 1 and 2)

3   Apply the principles developed in Block 6 to sketch the unit mesh of a given surface produced by either clean-surface or adsorbate-induced restructuring, and:

(a)   label the structure using the standard notation;

(b)   determine the fractional surface coverage of an adsorbate species. (SAQ 4)

4   For given examples of surface restructuring, apply the principles developed in Block 6 to:

(a)   interpret the changes that occur in LEED patterns;

(b)   interpret STM images.

(SAQ 3)

5   Outline the role that surface restructuring may play in dynamic (e.g. catalytic) processes, and comment on the suitability of surface techniques for the investigation of such dynamic processes.

SAQ ANSWERS AND COMMENTS

SAQ 1 (Objective 2)

The observed reversal of the interlayer relaxation results from the changes in bonding that occur with dissociative hydrogen adsorption. When hydrogen adsorbs on the surface layer, electrons on the rhodium atoms are used in bond formation to hydrogen. This withdraws electrons from the bonding between the top two layers of rhodium. These layers therefore move apart, tending towards the bulk spacing. Interactions between second and third layers, etc., respond similarly, all tending towards bulk-like behaviour.

SAQ 2 (Objective 2)

For *fcc* metals, the (111) surface is close packed, whereas the (210) surface is rather open or rough. We therefore expect significant vertical relaxation in the Pt(210) surface and little in the Ni(111) surface. In fact, the Ni(111) surface is bulk-truncated within the experimental accuracy of a LEED study ($\approx 2\%$), whereas relaxation in the Pt(210) surface has the largest-known value of 23% between layers 1 and 2.

SAQ 3 (Objective 4)

The simple LEED pattern (Figure 19b) obtained after oxygen treatment is of a (1×1) surface, the bulk-truncated structure of a (100) surface of a *fcc* metal. The spots, however, are not sharp, indicating some surface disorder and/or oxygen impurity. Heating in hydrogen removes the oxygen and leaves a clean well-ordered (1×1) surface, characterized by *sharp* spots.

SAQ 4 (Objective 3)

For the low-coverage hydrogen structure, the zigzag array of hydrogen atoms requires a (2×1) unit mesh, as shown in Figure 29a. According to the definition given in Section 7.3 of Block 6, the fractional surface coverage is $\theta = x/mn$, where x is the number of adsorbate species within the $(m \times n)$ unit mesh. In this case, $m = 2$, $n = 1$, and there are two hydrogen atoms in the unit mesh. So

$$\theta = \frac{2}{2 \times 1} = 1$$

For the high-coverage surface, the unit mesh is (1×2) as shown in Figure 29b. Thus, $m = 1$, $n = 2$, and there are three hydrogen atoms in the unit mesh. So

$$\theta = \frac{3}{1 \times 2} = 1.5$$

Like the example in SAQ 13 of Block 6, this is a situation in which the more precise definition of fractional surface coverage leads to a value of θ greater than one.

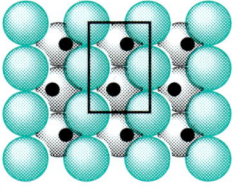

(a)

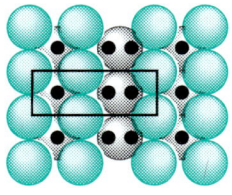

(b)

Figure 29   The surface unit meshes of the structures produced by hydrogen adsorption on Ni(110) at (a) low coverage, and (b) high coverage.

SAQ 5 (Objective 1)

Near the step edge, atoms migrate from *along* the edge, which is eaten away. Rows are formed by adding atoms to a flat terrace and not by row removal. This is an AR process (see Figure 30a).

On a terrace, the troughs that are created are not just a single row wide (which would lead to a MR structure). The material that is removed is deposited on the surrounding terrace, and so this too is an AR process (Figure 30b).

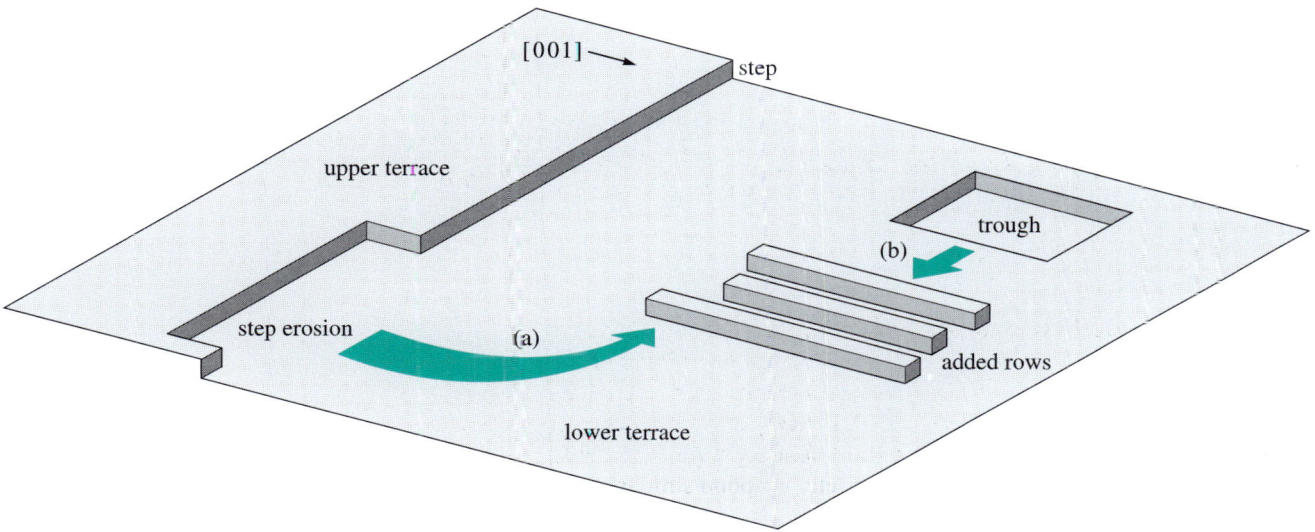

Figure 30   The formation of an added row (AR) structure by (a) step erosion, and (b) terrace erosion.

ACKNOWLEDGEMENTS

Grateful acknowledgement is made to the following sources for permission to reproduce material in this Topic Study:

Cover photograph: Lawrence Berkeley Laboratory/Science Photo Library; *Figure 2* reprinted from I. Stensgaard *et al.*, 'STM investigations of adsorption induced phases on metal surfaces', *Surface Science,* **269/270**, pp. 81–8, copyright 1992, with kind permission of Elsevier Science–NL, Sara Burgerhartstraat 25, 1055 KV, Amsterdam, The Netherlands; *Figure 4* reprinted from E. H. Conrad and T. Engel, 'The equilibrium crystal shape and the roughening transition on metal surfaces', *Surface Science,* **299/300**, p. 392, copyright 1994, with kind permission of Elsevier Science– NL, Sara Burgerhartstraat 25, 1055 KV, Amsterdam, The Netherlands; *Figure 11* D. King, 'Surface reconstructions', *Physics World,* **2**, March 1989, Institute of Physics Publishing Ltd; *Figure 12* reprinted from U. Köhler *et al.*, 'Scanning tunneling microscopy study of low-temperature epitaxial growth of silicon', *J. Vac. Sci. Technol. (A),* **7**, July/August 1989; *Figures 13 and 19* reprinted from K. Heinz, 'Geometrical and chemical restructuring of clean metal surfaces', *Surface Science,* **299/300**, pp. 439 and 441, copyright 1994, with kind permission of Elsevier Science– NL, Sara Burgerhartstraat 25, 1055 KV, Amsterdam, The Netherlands; *Figure 20a and b* reprinted from P. J. Estrup, 'Surface phases of reconstructed W(100) and Mo(100)', *Surface Science,* **299/300**, p. 727, copyright 1994, with kind permission of Elsevier Science–NL, Sara Burgerhartstraat 25, 1055 KV, Amsterdam, The Netherlands; *Figures 23 and 24* reprinted from F. Besenbacher et al., 'Chemisorption of H, O and S on Ni(110): general trends', *Surface Science,* **272**, pp. 335 and 336, copyright 1992, with kind permission of Elsevier Science–NL, Sara Burgerhartstraat